RECOVERING FROM CATASTROPHIC DISASTER IN ASIA

COMMUNITY, ENVIRONMENT AND DISASTER RISK MANAGEMENT VOLUME 18

RECOVERING FROM CATASTROPHIC DISASTER IN ASIA

EDITED BY

WILLIAM L. WAUGH, JR.
Georgia State University, USA

ZIQIANG HAN
Sichuan University, China

emerald
PUBLISHING

United Kingdom – North America – Japan
India – Malaysia – China

Emerald Publishing Limited
Howard House, Wagon Lane, Bingley BD16 1WA, UK

First edition 2017

British Library Cataloguing in Publication Data
A catalogue record for this book is available from the British Library

ISBN: 978-1-78635-296-5 (Print)
ISBN: 978-1-78635-295-8 (Online)
ISBN: 978-1-78714-893-2 (Epub)

ISSN: 2040-7262 (Series)

Printed and bound by CPI Group (UK) Ltd, Croydon, CR0 4YY

ISOQAR certified Management System, awarded to Emerald for adherence to Environmental standard ISO 14001:2004.

ISOQAR REGISTERED
Certificate Number 1985
ISO 14001

INVESTOR IN PEOPLE

CONTENTS

ABOUT THE EDITORS

William L. Waugh Jr. is professor emeritus in the Andrew Young School of Policy Studies at Georgia State University and the editor-in-chief of the *Journal of Emergency Management*. Dr. Waugh is the author of *International Terrorism* (1982); *Terrorism and Emergency Management* (1990); and *Living with Hazards, Dealing with Disasters* (2000); and the coeditor of *Handbook of Emergency Management* (1990); *Cities and Disasters* (1990); *Disaster Management in the U.S. and Canada* (1996); and *Emergency Management: Principles and Practice for Local Government*, 2nd ed. (2007); he has published over 200 articles, chapters, essays, and reports in the United States, Canada, Europe, and Asia. His current research focuses on collaboration in disaster management and capacity-building for community resilience.

Dr. Ziqiang Han is associate professor at the Institute for Disaster Management and Reconstruction, Sichuan University-The Hong Kong Polytechnic University in China. He is also a joint researcher at the Center for Crisis Management Research, Tsinghua University. His research interests are disaster recovery and sustainable development, multiorganizational cooperation in emergency management, risk perception and decision making, institutional disaster preparedness, and disaster education. Dr. Han has published a number of papers on *Natural Hazards; Risk, Hazards, and Crisis in Public Policy; International Journal of Disaster Risk Science; China Emergency Management Journal; Journal of Risk, Disaster and Crisis Research.*

ABOUT THE AUTHORS

Morio Onda is a professor of sociology at Ryutsu Keizai University, Chiba Prefecture, Japan. He received his Ph.D. in sociology from the University of Tokyo. He is the director of the Society of Economic Sociology and had edited the *Annual of the Society of Economic Sociology*. His research interests are the study of mutual help networks as spontaneous social order and indigenous social development. His publications include *Hatten no Keizaishakaigaku* (*Economic Sociology of Development*, Tokyo: Bunshindo, 1997), *Kaihatsushakaigaku* (*Development Sociology*, Kyoto: Minerva Publishing, 2001), *Gozyoshakairon* (*Mutual Help Society*, Kyoto: Sekaishisosha, 2006), and *Igakuseinotamenoshakaigaku* (*Sociology for Medical Students*, Kyoto: Koyoshobo, 2016).

José Manuel Mendes holds a Ph.D. in sociology from the Faculty of Economics of the University of Coimbra, Portugal, where he is an assistant professor with aggregation. He is also a researcher at the Centre for Social Studies, where he has been working in the fields of inequalities, social mobility, social movements, and collective action and, more recently, on the themes of risk and social vulnerability. He is the coordinator of the Risk Observatory (OSIRIS) of the Centre for Social Studies, co-coordinator of the Doctoral Programme "Territory, Risk and Public Policies," and is the editor-in-chief of *Revista Crítica de Ciências Sociais*. His most recent book is co-authored with Pedro Araújo: Sofrer e morrer onde se está (*To Suffer and to Die Where You Are*), Coimbra: Almedina, 2016.

David N. Nguyen was born and raised in Hawaii, United States. He received a master of arts in geography from the University of Hawaii, specializing on tourism, economics, and transportation geography. At the same time, he received a certificate in disaster management and humanitarian assistance from the Department of Urban and Regional Planning. He is currently a final year doctoral student at Tohoku University's Department of Civil Engineering, specializing in disaster management in the tourism sector and will graduate in 2017. During his time in Japan, David has worked in the Okinawan tourism industry, promoting tourism in the Tohoku Region of

Japan. David has published work related on sustainability, energy, international relations, and cuisine.

Fumihiko Imamura is a professor of tsunami engineering and the director of the International Research Institute of Disaster Science (IRIDeS) at Tohoku University. He received his doctoral of engineering in 1989 and became a research associate in 1989, an associate professor in 1992, and a professor of Tohoku University in 2000. He joined the IRIDeS in April 2012 as the vice director of the Disaster Control Research Center, Graduate School of Engineering, Tohoku University, in 2010–2012.

He is an expert on tsunami modeling, mitigation planning, and education/awareness. He has conducted field surveys as leader for earthquakes and tsunami damage investigation since the 1992 Nicaragua and Flores Island in Indonesia. In addition, he is the secretary for the International TIME (Tsunami Inundation Modeling Exchange) Project, which is supported by the Intergovernmental Oceanographic Commission, the International Union of Geodesy, and the Geophysics Tsunami Commission.

Chandra Lal Pandey is an assistant professor at the School of Arts, Kathmandu University. He earned his Ph.D. and postdoctoral fellowship from the University of Waikato, New Zealand. His research expertise includes environmental politics and policy, natural resource and crisis management, community development, resilience building, and crisis management. He has published a number of research articles in high-quality journals including *Asian Security and Environmental Values*. His most recent publication includes a co-edited book *Environmental Security in the Asia-Pacific*, a co-authored book *Why REDD Will Fail*, and a journal article "Managing Climate Change: Shifting roles for NGOs."

Yingying Sun is an associate professor in Information at the Institute for Disaster Management and Reconstruction (IDMR), Sichuan University-The Hong Kong Polytechnic University. Dr. Sun's research interests focus on the following methods and disciplines: action research, narrative, semistructured interview methods; understanding problems of minority people in disaster risk mitigation; qualitative analysis through ethnographic and psychological theories. Dr. Sun continues to do action research to promote risk awareness in communities and to develop disaster education tools at schools. Dr. Sun's study includes disaster risk mitigation, response and recovery based on research in Japan (earthquake, tsunami, and flood), Mexico (earthquake and tsunami), China (earthquake, landslide, and flood), Taiwan (earthquake and typhoon), and Australia (cyclone and flood).

Shinya Uekusa is currently a Ph.D. candidate at the University of Auckland (Department of Sociology) in New Zealand. His main research interest is in social vulnerability and resilience in disasters, focusing on immigrants, refugees, and linguistic minorities in Canterbury, New Zealand and Tohoku, Japan.

Takashi Tsuji is a junior research associate at the Regional Environmental Renovation Section, Fukushima Branch, National Institute for Environmental Studies, Japan. He is a Ph.D. candidate at the Nagoya University, Japan. His research interests are disaster governance, community resilience after disaster, focusing on long-term community development process.

LIST OF CONTRIBUTORS

Ziqiang Han	Institute for Disaster Management and Reconstruction, Sichuan University-The Hong Kong Polytechnic University, Chengdu, China
William L. Waugh, Jr.	Georgia State University, Decatur, GA, USA
José Manuel Mendes	Centre for Social Studies, Faculty of Economics, University of Coimbra, Coimbra, Portugal
David N. Nguyen	Graduate School of Civil Engineering, Tohoku University, Sendai, Japan
Fumihiko Imamura	International Research Institute of Disaster Science, Tohoku University, Sendai, Japan
Morio Onda	Department of Sociology, Ryutsu Keizai University, Matsudo, Japan
Chandra Lal Pandey	School of Arts, Kathmandu University, Hattiban, Lalitpur, Nepal
Yingying Sun	Institute for Disaster Management and Reconstruction, Sichuan University-The Hong Kong Polytechnic University, Chengdu, China
Takashi Tsuji	National Institute for Environmental Studies, Fukushima, Japan
Shinya Uekusa	Department of Sociology, University of Auckland, Auckland, New Zealand

CHAPTER 1

DISASTER RECOVERY IN ASIA: AN INTRODUCTION

Ziqiang Han and William L. Waugh, Jr.

ABSTRACT

This chapter provides the foundation for the book. The objective of this chapter is to outline the theme of the book and to provide the context for the chapters that follow. Disaster recovery is a challenge for governments and for affected communities, families, and individuals. It is a challenge, because recovery from catastrophic disasters can be much more complicated and elusive than what can be addressed by national and international aid organizations given the time and other resources. The short literature review provides the research context, and the overview of the book describes each of the chapters briefly.

Keywords: Recovery; disaster; literature review

Asia has experienced devastating disasters over the centuries. Proximity to the seismically active "Ring of Fire" and other plate boundaries, long Pacific and Indian Ocean coastlines, major river and tributary courses, desert and semidesert areas, and other geographic features create a diversity of risks and potential hazards. The very size of the Asian continent means that hazards differ

Recovering from Catastrophic Disaster in Asia
Community, Environment and Disaster Risk Management, Volume 18, 1–14

ISSN: 2040-7262/doi:10.1108/S2040-726220160000018001

considerably from one region to another. It also means that it is difficult to generalize about risk and vulnerabilities in Asia. However, history has demonstrated that the risk of catastrophic disaster is very real. One listing of "worst" natural disasters included the 1876–1979 Famine in North China, which left 9 million dead, and the 1931 Yellow River Floods in Central China, which left 4 million dead. River flooding, cyclones, and earthquakes with hundreds of thousands dead completed the list (Szczepanski, 2017). Greater attention is also being paid to the slow onset disasters associated with climate change.

John West (2014), the director of the Asian Century Institute, has stated that "Asia is the world's most disaster-prone region…." According to the Emergency Events Database (EM-DAT) that collects disaster impact data globally since 1988, Asia is the most frequently impacted continent, having 9,380 occurrences of disasters during 1900–2016, and sharing about 41.65% of the global events. In the last decade (2004–2014), 80% of disaster fatalities in the world occurred in Asia mostly due to earthquakes and tsunamis. Drought and famine also caused a substantial percentage of fatalities. The decade, cited by the Asian Century Institute, included the 2004 Indian Ocean earthquake and tsunami with 260,000 deaths, the 2005 Pakistan earthquakes with 75,000 deaths, the 2008 Sichuan earthquake with 90,000 deaths, the 2008 Myanmar Cyclone Nargis disaster with 140,000 deaths, and Japan's 2011 earthquake, tsunami, and nuclear facility failure. Floods, droughts, severe storms, and other events associated with climate change contributed to the death toll in Asia. In 2008 alone, humanitarian aid totaled US$16.9 billion with most going for short-term emergency response and relief (West, 2014). Note that the human costs of disasters vary by source and sometimes are unknown.

Changes in response and recovery approaches vary over time, and some of the 2004–2014 disasters have had significant impact on how we respond to disasters. For example, the 2004 Indian Ocean earthquake and tsunami provided lessons on the management of mass casualty events. So many people were killed, and identifying casualties and preserving records were difficult. The tsunami was the largest disaster in Swedish history because of the number of Swedish nationals who lost their lives in Thailand due to the tsunami (Elhakeem, 2014). Considerable attention has been paid to the 2005 Pakistan earthquakes because of the damage done to infrastructure, as well as to people. School buildings collapsed. Roads were destroyed in areas already difficult to access. Water reservoirs were destroyed, forcing a return to traditional, less safe, and less reliable water sources. The 2011 Japan earthquake and tsunami became a cascading event in which the earthquake created a tsunami and damaged a nuclear facility. Disasters are often cascading events, and we learn to anticipate the impacts of disaster, such as the fires that are caused by earthquakes and wildfires caused by drought and heat waves.

Why does Asia have so many catastrophic disasters? High population density, extreme poverty, and a geography that leaves many communities vulnerable to hazards contribute to the risk. Weak governance structures and economies in some large countries complicate response and recovery. In addition to human casualties, homes are lost, businesses are destroyed and the means of production are interrupted, infrastructure is destroyed and access is obstructed, farm land is lost and food production is interrupted, residents are displaced, and the environment is so severely damaged that affected populations do not want to return to their homes. Rebuilding lives is not the only challenge. Some communities do not recover while some communities recover remarkably quickly. Not only the scale of the disaster but also many other variables affect the recovery process. Local leadership (governance) and internal and external linkages (social capital) are also important (Aldrich, 2017; Alesch, Arendt, & Holly, 2009; Dynes, 2006; Phillips, 2009).

Recounting the details of thousands of catastrophic disasters in Asia would be a monumental task. In 2015, there were US$45.1 billion in economic damage, 53.9 million people affected, and 16,046 deaths, in 160 natural disasters. The disasters included earthquakes in Nepal, Afghanistan, Tajikistan; 33 major tropical cyclones in Southeast Asia and in the Pacific; severe flooding in China, India, Indonesia, Myanmar, Pakistan, and Sri Lanka; deadly heat waves in India and Pakistan; and major droughts covering large areas of the continent. In 2015, two major international agreements were adopted to increase sustainable development and to address the impacts of climate change and increase resilience. Areas identified as needing attention were building resilience in urban areas, increasing regional cooperation, addressing slow-onset disasters, early warning systems, and use of technology in disaster response and recovery (United Nations Economic and Social Commission for Asia and the Pacific, 2016).

Linking disasters and development is challenging, particularly for the governments that have to align policies and programs and resources to make the linkage work (Kapucu & Liou, 2014). The breadth of disaster recovery issues makes policy making, program design, and program implementation very difficult. Successful linkage requires considerable investment in political and administrative capacity building (Waugh & Liu, 2014). Central government agencies cannot always provide housing that is appropriate for the culture (Ganapati, 2014). Disaster recovery involves social, economic, political, cultural, and physical processes in cultural contexts that are often very diverse. Rebuilding communities has to include rebuilding social networks, restoring livelihoods, improving governance capacities, and supporting local culture and values, as well as reconstruction of homes and businesses and infrastructure. Rebuilding lives is a much more complicated process than rebuilding structures. Moreover, linking disasters to development is the "building back better" approach that the international community supports, and national governments are encouraged to adopt conflicts with pressures, to act quickly with medical, food, and shelter aid, and to stay engaged with affected communities for an extended period. Common wisdom is to focus on building community resilience before and during disasters so that communities can manage their own recovery or communicate their priorities to regional and central authorities and to external aid missions. Greater resilience can speed recovery and ensure that communities are better prepared for the next natural or man-made disaster. Greater resilience can also help affected communities feel less as victims and more as citizens in control of their own destinies.

Both practitioners and researchers recognize that disaster recovery is the least understood aspect of emergency management (Smith & Wenger, 2007). Claire Rubin reviewed her 30 years of experience and studies on long-term recovery in 2009, and she concluded, "the research and knowledge base in the realm of long-term recovery is seriously inadequate to the needs we face today" (Rubin, 2009, p. 1). Moreover, most of these studies mainly reflected experiences from the Western countries, the United States in particular, and there is a lack of Asia's perspective.

Restoration, reconstruction, rehabilitation, and reinstitution are similar terms that are often used interchangeably with disaster recovery. Restoration implies that things are brought back to the original pre-disaster state after a disaster, whereas rehabilitation focuses more on the restoration of people than things. Similarly, reinstitution suggests some kind of restoration of the rightful claimants of owners. Reconstruction, on the other hand, centers on the physical rebuilding of human communities in the post-disaster period. Recovery, the most inclusive term, refers to "moving a disaster-impacted community to a healthy state which can include restoration, reconstruction and social change," which "may or may not be the same as the pre-impact level" (Dynes & Quarantelli, 2008; Quarantelli, 1999).

Early disaster recovery studies recognized recovery as ordered, knowable, and predicable, with an emphasis on the building environment (Haas, Kates, & Bowden, 1977). For example, in Haas et al.'s classic study, the recovery process consisted four overlapping periods: the emergency period, the restoration period, the replacemental reconstruction period, and the commemorative, betterment, and developmental period. However, later studies have shown that the recovery process does not follow a predictable timeline and that the recovery process is actually a dynamic, interactive decision-making process rather than a static and linear process (Mileti, 1999; Nigg, 1995; Rubin and

Popkin, 1990). Disaster recovery could be conceptualized as "the differential process of restoring, rebuilding, and reshaping the physical, social economic and natural environment through pre-event planning and post-event actions" (Smith & Wenger, 2007, p. 237).

One question always raised in disaster recovery studies is recovery to where, that is, the goals of recovery. Generally, there are three categories of recovery goals: (1) the restoration of the status quo to the pre-disaster situation, (2) reconstruction as a chance to realize structural improvement, and (3) no clearly defined recovery goals. Of course, there is usually the simultaneous co-existence of multiple recovery aims in reality, and the goals may change periodically (Geipel, 1991). The approach of perceiving "recovery" as restoration can be problematic, because returning to pre-disaster levels does not necessarily mean building back for the better (Ganapati, Cheng, & Ganapati, 2012). Furthermore, recovering to the pre-disaster situation implies restoring the pre-event inequality, exploitation, and vulnerability as well (Oliver-Smith, 1990). This is especially common in some underdeveloped areas in developing countries with extreme poverty, chronic injustice and exploitation, and high-risk exposure, as has been witnessed in recent disasters in Pakistan (Mustafa, 2003) or Haiti (Olshansky & Etienne, 2011). The idea of building back better or recovering better should be adopted, especially in the case of developing countries where building back better is indeed possible (Mulligan & Nadarajah, 2012) if the ideas of development, vulnerability, and risk reduction are integrated into recovery activities (Shaw, 2006), with the physical and social planning integrated with one another to address local needs in culturally appropriate ways (Mulligan, Ahmed, Shaw, Mercer, & Nadarajah, 2012). Hence, beyond a return to previous "normality," there are opportunities to integrate disaster mitigation, vulnerability reduction, and sustainable development into the recovery process, in hopes of achieving improvement through recovery.

In terms of *measurement*, most of studies on family recovery tend to adopt a multidimensional measurement. Housing and economy are the two commonly included dimensions from the physical aspect, while individuals' perception of recovery and/or social–psychological quality of life are other dimensions. Social role adaptation and physical health recovery are the ones less included. In Haas et al.'s classic study about family recovery, housing recovery, jobs recovery, and perceived recovery were used as the indicators of family recovery. Following the two crosscutting themes of the importance of kinship linkage and the loss of family functions in early disaster studies, long-term family recovery in Bolin's work predominately covers three dimensions: housing recovery, economic recovery, and emotional recovery in terms of quality of life (Bolin, 1982, 1994). Bates and Peacock developed the Domestic Assets Scale, which captures the economic value of household facilities of household functional areas (Bates & Peacock, 1992). It was later modified and adopted for assessing the impact of Indian Ocean tsunamis on households (Arlikatti, Peacock, Prater, Grover, & Sekar, 2010). Recent studies about disaster recovery include more dimensions. For instance, Abramson et al. developed a five-dimensional measurement for individual disaster recovery after the Hurricane Katrina disaster, including housing stability, economic stability, physical health, mental health, and social role adaptation (Abramson, Stehling-Ariza, Park, Walsh, & Culp, 2010). Another study examining disaster survivors' recovery after the Bam earthquake in Iran employed a measure of quality of life that included four dimensions: physical health, psychological state, social relationships, and environment (Ardalan et al., 2011). In Tatsuki (2007)'s study about life recovery after the 1995 Kobe earthquake, life recovery measure was drawn from 14 items that captured the respondents' perception of life fulfillment/readjustment (seven items), life satisfactions (six items), and future prospect (one item). Han, Ba, Xin, and Zhong (2016) adopted the Sustainable Livelihoods Approach's physical, human, natural, financial, and social capital's measurement in their study on family recovery after the 2008 Wenchuan earthquake in China.

For *measurements at the community or macro level*, housing and population restoration are the two most common dimensions included. However, many studies included more dimensions, such as physical infrastructure or social infrastructure like community linkages. Alesch et al. (2009) proposed that community disaster recovery should include the restoring of basic services, replacing damaged infrastructure capacity, rebuilding critical social and economic elements of the community system, or reestablishing relationships among critical elements of the community. A wide range of indicators, such as changes in population and residential units, vacancy rates, affordability of housing, retention of local residents, structural improvements, extent of retrofitting, quality of life, the time taken for reconstruction, the quality of reconstruction, and residents' satisfaction, are proposed to measure recovery outcomes in some studies (Loukaitou-Sideris & Kamel, 2004). On the basis of recovery experience in New Orleans after Hurricane Katrina, housing rehabilitation, public service and infrastructure, and labor force and employment are recognized as the key indicators of recovery at the macro level for analysis (Liu, Fellowes, Mabanta, & Program, 2006). In the special issue of the *International Journal of Mass Emergencies and Disasters* in 2012, which covered the theorization efforts of disaster recovery, the built environment (Alesch & Siembieda, 2012), ecosystems (Berke & Glavovic, 2012), economic recovery (Chang & Rose, 2012), institutional dimensions (Smith & Birkland, 2012), and social dimensions (Tierney & Oliver-Smith, 2012) of disaster recovery are discussed.

In *determinants of family or household recovery*, there are five major clusters of influencing factors: the socioeconomic characteristics of the family; the external aid coming from informal personal social networks and local organizations; disaster impact and disruption degrees; macro community features such as pre-disaster planning, post-event response, and recovery, as well as the collective activities within communities; and external institutional help such as higher levels of government assistance programs. In early studies of family recovery, the family is always seen as an open system that can use both internal and external resources. The demographic attributes, disaster impact (including both direct impact and social disruption), and family recovery capacity, including both internal resources like insurance or financial assets or external resources through their personal social network or aid from formal organizations like government or nongovernmental organizations are the main factors influencing family recovery after disasters (Bolin, 1994; Drabek & Key, 1984; Haas et al., 1977). For the demographic attributes, most studies found that the disaster recovery status would vary according to the main family members' life cycle (age) and ethnicity and that socioeconomic characteristics are measured in varied ways such as income or saving (Arlikatti & Andrew, 2011; Bolin, 1994; Drabek & Key, 1984; Haas et al., 1977; Peacock, Killian, & Bates, 1987). These internal factors may directly or indirectly influence the external resources and paths of getting resources.

Determinants of community recovery or recovery at larger area: Quarantelli and Dynes have summarized the determinants of disaster recovery (Dynes & Quarantelli, 2008; Quarantelli, 1999). According to their summary, predisaster patterns and social trends, economic factors, government policies, and prior community recovery planning were the main influencing factors involved in disaster recovery. From the institutional perspective, effective intergovernmental relationships are essential to efficient recovery. The long-term recovery process is a dynamic process that involved both federal and state influence, community needs for action, and community planning and implementation capacity, including personal leadership and the ability to act and the knowledge of what to do (Rubin, 1985; Rubin & Barbee, 1985).

Federal Emergency Management Agency (FEMA) summarized the common features among communities that had successful recovery stories from their long-term community recovery programs. Acting quickly; planning for recovery proactively; engaging the community; developing partnerships, networks, and effective coordination strategies; making decisions and managing recovery locally; managing financial acquisition well; keeping organizations flexible; integrating mitigation; and preparedness into recovery are the common features of community with better recovery (FEMA, 2011a, 2011b). The Government Accountability Office (GAO) also concluded that (1) having clearly defined recovery roles and responsibilities among stakeholders, (2) effective coordination and collaboration among recovery stakeholders, and (3) periodic evaluation of and reporting on the recovery process were the three primary characteristics of successful disaster recovery efforts after reviewing five catastrophic disasters (Czerwinski, 2009, 2012). Lessons from developing countries also suggested that (1) incorporating long-term recovery goals into disaster response and pre-disaster planning, (2) expanding the knowledge base by incorporating research into recovery and harnessing lessons learned from international experiences, and (3) developing an outcome-oriented approach to disaster recovery planning, including the measurement of community-level outcomes, could be key approaches to enhance disaster recovery (Garnett & Moore, 2010).

The importance of personal network in facilitating family recovery and the collective actions within community concepts are reconstructed using the term of social capital in recent years (Aldrich, 2012; Dynes, 2006). Studies examining the role of social capital, regardless of measured from the network approach or civic engagement method, in either micro or macro levels of recovery indicate that social capital plays an important but complex role in disaster recovery. First, social capital, especially the resources embedded in personal social networks are important for both response and recovery (Fussell, 2006). Second, the feeling of bonding and cohesion within communities could facilitate collective action for post-disaster recovery (Chamlee-Wright & Storr, 2011a; Storr & Haeffele-Balch, 2012). However, it may also strengthen the obstacles for people on the periphery of society by pushing the "public bad" to other communities (Aldrich & Crook, 2008), increase the possibility of rent seeking for available resources (Chamlee-Wright & Storr, 2011b), or push the social capital disadvantaged groups to a more vulnerable situation (Hawkins & Maurer, 2010), especially when resources were limited.

OVERVIEW OF THIS BOOK

Chapter 2, "Social Capital and Changes in Post-disaster Recovery Process: Observations From China After the 2008 Wenchuan Earthquake," is written by Ziqiang Han of the Institute for Disaster Management and Reconstruction, Sichuan University–The Hong Kong Polytechnic University. Using data from a two-wave survey of households affected by the earthquake, he concludes that informal organizations were the most prominent, formal organizations that had the greater impact. The average degree of social capital increased during the recovery, and inequalities in social capital decreased. Recovery can strengthen communities.

Chapter 3, "Recovering From Prolonged Negative Destination Images in Post-disaster Northern Japan," is written by David N. Nguyen, a Ph.D. scholar in the Graduate School of Civil Engineering, Tohoku University, and Fumihiko Imamura, the Director of the International Research Institute of Disaster Science, Tohoku University, Japan. They focus on the impact of the 2011 Great East Japan Earthquake and Tsunami on tourism in the Tohoku region. Despite the rebuilding of infrastructure, tourism in the region has not returned to pre-disaster levels. They examine the different media strategies adopted by local officials to change the images of their communities. They recommend more research to determine how best to change tourists' risk perceptions.

Chapter 4, "Restoration of Communities Following the Great East Japan Disaster: The Transformation of Mutual Help Networks Through the Eyes of the Victims," by Morio Onda from Ryutsu Keizai University in Japan, focuses on the perspectives of the victims of the 2011 Great East Japan Earthquake and Tsunami. Using interviews of disaster victims, Onda explores the restoration of social bonds and a sense of community and the development of mutual aid networks. One conclusion is that the greater the self-reliance, the greater the social bonds. The authors also conclude that outside assistance reduces internal connections. Recommendations for further research are included.

Chapter 5, "Lessons From Disaster Recovery in Japan Through Case Studies of Four Earthquakes," is written by Yingying Sun from the Institute for Disaster Management and Reconstruction (IDMR), Sichuan University, Chengdu, China. Sun examines the characteristics of natural disasters in Japan to draw insights on domestic and international disaster response and recovery efforts. Recovery experiences following the Great Hanshin Earthquake of 1995, the West Tottori Earthquake in 2000, the Niigata Chuetsu Earthquake in 2004, and the Tohoku Earthquake and Tsunami of 2011 are compared. Lessons are drawn as Japan prepares for an anticipated major earthquake in the Nankai trough in the next 30 years.

Chapter 6, "Citizen Participation in the Disaster Reconstruction Process: Lessons From the Great East Japan Earthquake," is written by Takashi Tsuji from the National Institute for Environmental Studies in Japan. Tsuji focuses on citizen participation in reconstruction following the earthquake through the development of business plans developed by local governments. On the basis of field work in three affected communities, Tsuji examines how residents were organized, how community organizations contributed to the participation, and how governance and community organization need to change to ensure local representation and the political legitimacy of decisions.

Chapter 7, "Social Vulnerability in Disasters: Immigrant and Refugee Experiences in Canterbury and Tohoku," is written by Shinya Uekusa from the Department of Sociology, University of Auckland, New Zealand. Uekusa compares the experiences of linguistic minority immigrants and refugees in the two disasters. Data from interviews and two datasets were used to see whether the two groups were more vulnerable than other affected populations. Uekusa concludes that the two groups may be more resilient because of their past experiences with war and other hardships and challenges common wisdom concerning social vulnerability.

Chapter 8, "Disaster Exceptionalism in India: The View From Below," is written by José Manuel Mendes from the Centre for Social Studies, Faculty of Economics, University of Coimbra, Portugal. He focuses on the Kosi River floods in the state of Bihar in August 2008 and its impact on Dalit communities. Using interview data involving national and local leaders and activists in NGOs and Dalit organizations, he concludes that the top-down approach of Indian states under the Disaster Management Act of 2005 gives the state authority to dominate the response and recovery processes, thus permitting social and economic inequities to influence the delivery of assistance.

Chapter 9, "The 2015 Nepal Earthquake: From Rescue to Reconstruction," is written by Chandra Lal Pandey from Kathmandu University, Institute of Crisis Management Studies at Tribhuvan University, and the Southasia Institute of Advanced Studies, Kathmandu, Nepal. He focuses on the impact of the 2015 Nepal Earthquakes that affected large areas of the country and the current status of the recovery effort. He points out information gaps and makes recommendations for further research to improve disaster response and recovery.

Chapter 10, "International Humanitarian Assistance and Disaster Recovery in Asia," is written by William L. Waugh, Jr., Professor Emeritus, Andrew Young School of Policy Studies, Georgia State University in Atlanta. He examines the role of humanitarian assistance in catastrophic disasters and concludes that, despite the development of international agreements, such as the Hyogo and Sendai Frameworks, and international standards for humanitarian assistance, it is difficult for international aid organizations to find the funding and time to engage with affected communities and remain on site long enough to sustain long-term recovery programs. Aid missions generally arrive quickly and work until the available funding ends, and they may not have time to work with local officials and residents to assure that their priorities are met. Experienced outside agencies may help focus attention on issues like corruption, human trafficking, and gender-related abuse, however.

The collection is multidisciplinary and the perspectives and methodologies are diverse. Each chapter addresses critical issues in catastrophic disaster recovery from politics to restoring the social, cultural, and economic lives of affected populations.

REFERENCES

Abramson, D. M., Stehling-Ariza, T., Park, Y. S., Walsh, L., & Culp, D. (2010). Measuring individual disaster recovery: A socioecological framework. *Disaster Medicine and Public Health Preparedness*, *4*, S46–S54.

Aldrich, D. J. (2012). *Building resilience: Social capital in post-disaster recovery*. Chicago, IL: University of Chicago Press.

Aldrich, D. P. (2017). The importance of social capital in building community resilience. In W. Yan & W. Galloway (Eds.), *Rethinking resilience, adaptation and transformation in a time of change* (pp. 357–364). New YorK: Springer International Publishing. Retrieved from https://doi.org/10.1007/978-3-319-50171-0_23.

Aldrich, D. P., & Crook, K. (2008). Strong civil society as a double-edged sword: Siting trailers in Post-Katrina New Orleans. *Political Research Quarterly*, *61*(3), 379–389.

Alesch, D. J., Arendt, L. A., & Holly, J. N. (2009). *Managing for long-term community recovery in the aftermath of disaster*. Fairfax, VA: Public Entity Risk Institute.

Alesch, D. J., & Siembieda, W. (2012). The role of the built environment in the recovery of cities and communities from extreme events. *International Journal of Mass Emergencies and Disasters, 30*(2), 197–211.

Ardalan, A., Mazaheri, M., Vanrooyen, M., Mowafi, H., Nedjat, S., Naieni, K. H., et al. (2011). Post-disaster quality of life among older survivors five years after the Bam earthquake: Implications for recovery policy. *Ageing and Society*, *31*(02), 179–196.

Arlikatti, S., & Andrew, S. (2011). Housing design and long-term recovery processes in the aftermath of the 2004 Indian Ocean tsunami. *Natural Hazards Review*, *13*(1), 34–44.

Arlikatti, S., Peacock, W. G., Prater, C. S., Grover, H., & Sekar, A. S. G. (2010). Assessing the impact of the Indian Ocean tsunami on households: A modified domestic assets index approach. *Disasters*, *34*(3), 705–731.

Bates, F. L., & Peacock, W. G. (1992). Measuring disaster impact on household living conditions: the domestic assets approach. *International Journal of Mass Emergencies and Disasters*, *10*(1), 133–160.

Berke, P. R., & Glavovic, B. C. (2012). Ecosystems and disaster resiliency: Contributions to a holistic theory of recovery. *International Journal of Mass Emergencies and Disasters*, *30*(2), 182–196.

Bolin, R. C. (1982). *Long-term family recovery from disaster. Boulder*, CO: Institute of Behavioral Science, University of Colorado Boulder.

Bolin, R. C. (1994). *Household and community recovery after earthquakes. Boulder*, CO: Institute of Behavioral Science, University of Colorado Boulder.

Chamlee-Wright, E., & Storr, V. H. (2011a). Social capital as collective narratives and post-disaster community recovery. *Sociological Review*, *59*(2), 266–282.

Chamlee-Wright, E., & Storr, V. H. (2011b). Social capital, lobbying and community-based interest groups. *Public Choice*, *149*(1–2), 167.

Chang, S., & Rose, A. (2012). Towards a theory of economic recovery from disasters. *International Journal of Mass Emergencies and Disasters*, *30*(2), 171–181.

Czerwinski, S. J. (2009). *Disaster recovery: Experiences from past disasters offer insights for effective collaboration after catastrophic events*. Retrieved from http://www.gao.gov/products/GAO-09-811

Czerwinski, S. J. (2012). *Disaster recovery: Selected themes for effective long-term recovery*. Retrieved from http://www.gao.gov/products/GAO-12-813T.

Drabek, T. E., & Key, W. H. (1984). *Conquering disaster: Family recovery and long-term consequences*. New York, NY: Irvington Publishers.

Dynes, R. R. (2006). Social capital: Dealing with community emergencies. *Homeland Security Affairs*, *2*(2), 1–26.

Dynes, R. R., & Quarantelli, E. L. (2008). *A brief note on disaster restoration, reconstruction and recovery: A comparative note using post earthquake observations*. Disaster Research Center, University of Delaware, Newark, DE.

Elhakeem, N. (2014). The wave Sweden will not forget. *The Local*, 22 December.

FEMA. (2011a). *Lessons in community recovery: Seven years of emergency support function #14 long-term community recovery from 2004 to 2011*. Retrieved from https://www.fema.gov/pdf/rebuild/ltrc/2011_report.pdf.

FEMA. (2011b). *National disaster recovery framework: Strengthening disaster recovery for the nation*. Retrieved from https://www.fema.gov/pdf/recoveryframework/ndrf.pdf.

Fussell, E. (2006). Leaving New Orleans: Social stratification, networks, and hurricane evacuation. *Understanding Katrina: Perspectives from the social sciences*. New York, NY: Social Sciences Research Council.

Ganapati, N. E. (2014). Linking Development to disasters in Turkey: Moving Forward after the Marmara Earthquake. In N. Kapucu & K. T. Liou (Eds.), *Disaster and Development* (pp. 61–78). New York: Springer International Publishing. Retrieved from https://doi.org/10.1007/978-3-319-04468-2_4.

Ganapati, N. E., Cheng, S., & Ganapati, S. (2012). Resilient rural communities? Housing recovery patterns following Hurricane Katrina. In N. Kapucu, C. V. Hawkins, & F. I. Rivera (Eds.), *Disaster resiliency: Interdisciplinary perspectives*. New York: Taylor & Francis.

Garnett, J. D., & Moore, M. (2010). Enhancing disaster recovery: Lessons from exemplary international disaster management practices. *Journal of Homeland Security and Emergency Management*, *7*(1), 1–20.

Geipel, R. (1991). *Long-term consequences of disasters: The reconstruction of Friuli, Italy, in its international context, 1976–1988*: New York: Springer.

Haas, J. E., Kates, R. W., & Bowden, M. J. (1977). *Reconstruction following disaster.* Cambridge, MA: MIT Press.

Han, Z., Ba, Z., Xin, R., & Zhong, P. (2016). Rural households recovery after disasters: From the sustainable livelihoods perspective. *China Population, Resources and Environment*, *26*(4), 158–167. [韩自强, 巴战龙, 辛瑞萍, & 钟平. (2016). 基于可持续生计的农村家庭灾后恢复研究. 中国人口资源与环境, *26*(4), 158–167.]

Hawkins, R. L., & Maurer, K. (2010). Bonding, bridging and linking: How social capital operated in New Orleans following Hurricane Katrina. *British Journal of Social Work*, *40*(6), 1777–1793.

Kapucu, N., & Liou, K. T. (2014). Disasters and Development: Investigating an Integrated Framework. In N. Kapucu & K. T. Liou (Eds.), *Disaster and development* (pp. 1–15). New York: Springer International Publishing. https://doi.org/10.1007/978-3-319-04468-2_1.

Liu, A., Fellowes, M., Mabanta, M., & Program, B. I. M. P. (2006). *Special edition of the Katrina Index: A one year review of key indicators of recovery in post-storm New Orleans*: Brookings Institution.

Loukaitou-Sideris, A., & Kamel, N. M. (2004). *Residential recovery from the Northridge earthquake: An evaluation of federal assistance programs*: Policy Research Program, California Policy Research Center.

Mileti, D. (1999). *Disasters by design: A reassessment of natural disasters in the United States*. Washington, D.C.: Joseph Henry Press.

Mulligan, M., & Nadarajah, Y. (2012). Rebuilding community in the wake of disaster: Lessons from the recovery from the 2004 tsunami in Sri Lanka and India. *Community Development Journal, 47*(3), 353–368.

Mulligan, M., Ahmed, I., Shaw, J., Mercer, D., & Nadarajah, Y. (2012). Lessons for long-term social recovery following the 2004 tsunami: Community, livelihoods, tourism and housing. *Environmental Hazards-Human and Policy Dimensions, 11*(1), 38–51. doi:10.1080/17477891.2011.635186

Mustafa, D. (2003). Reinforcing vulnerability? Disaster relief, recovery, and response to the 2001 flood in Rawalpindi, Pakistan. *Global Environmental Change Part B: Environmental Hazards, 5*(3), 71–82.

Nigg, J. M. (1995). *Disaster recovery as a social process*. Disaster Research Center, University of Delaware, Newark, DE.

Oliver-Smith, A. (1990). Post-disaster housing reconstruction and social inequality: A challenge to policy and practice. *Disasters, 14*(1), 7–19.

Olshansky, R. B., & Etienne, H. F. (2011). Setting the stage for long-term recovery in Haiti. *Earthquake Spectra, 27*(S1), 463–486.

Peacock, W. G., Killian, C. D., & Bates, F. L. (1987). The effects of disaster damage and housing aid on household recovery following the 1976 Guatemalan earthquake. *International Journal of Mass Emergencies and Disasters, 5*(1), 63–88.

Phillips, B. D. (2009). *Disaster recovery*. Boca Raton, FL: CRC Press.

Quarantelli, E. L. (1999). *The disaster recovery process: What we know and do not know from research*. Disaster Research Center, University of Delaware, Newark, DE.

Rubin, C. B. (1985). The community recovery process in the United States after a major natural disaster. *International Journal of Mass Emergencies and Disasters, 3*(2), 9–28.

Rubin, C. B. (2009). Long term recovery from disasters—The neglected component of emergency management. *Journal of Homeland Security and Emergency Management, 6*(1), 1–17.

Rubin, C. B., & Barbee, D. G. (1985). Disaster recovery and hazard mitigation: Bridging the intergovernmental gap. *Public Administration Review, 45*, 57–63.

Rubin, C. B., & Popkin, R. (1990). *Disaster recovery after hurricane Hugo in South Carolina*. Natural Hazards Research and Applications Information Center, Institute of Behavioral Science, University of Colorado, Boulder, CO.

Shaw, R. (2006). Indian Ocean tsunami and aftermath: Need for environment-disaster synergy in the reconstruction process. *Disaster Prevention and Management, 15*(1), 5–20.

Smith, G., & Birkland, T. (2012). Building a theory of recovery: Institutional dimensions. *International Journal of Mass Emergencies and Disasters, 30*(2), 147–170.

Smith, G. P., & Wenger, D. (2007). Sustainable disaster recovery: Operationalizing an existing agenda. In *Handbook of disaster research* (pp. 234–257). New York: Springer. Retrieved from https://doi.org/10.1007/978-0-387-32353-4_14.

Storr, V. H., & Haeffele-Balch, S. (2012). Post-disaster community recovery in heterogeneous, loosely connected communities. *Review of Social Economy*, *70*(3), 295–314.

Szczepanski, K. (2017). *Asia's worst natural disasters*. Retrieved from Thoughtco.com Humanities, https://www.thoughtco.com/asias-worst-natural-disasters-195150. Accessed 9 May 2017.

Tatsuki, S. (2007). Long-term life recovery processes among survivors of the 1995 Kobe earthquake: 1999, 2001, 2003, and 2005 Life Recovery Social Survey results. *Journal of Disaster Research*, *2*(6), 484–501.

Tierney, K., & Oliver-Smith, A. (2012). Social dimensions of disaster recovery. *International Journal of Mass Emergencies and Disasters*, *30*(2), 123–146.

United Nations Economic and Social Commission for Asia and the Pacific. (2016). *Disasters in Asia and the Pacific: 2015 year in review*. Retrieved from http://www.unescap.org/sites/default/files/2015_Year%20in%20Review_final_PDF_1.pdf. Accessed 9 May 2017.

Waugh, W. L., & Liu, C. Y. (2014). Disasters, the Whole community, and development as capacity building. In N. Kapucu & K. T. Liou (Eds.), *Disaster and development* (pp. 167–179). New York: Springer International Publishing. Retrieved from https://doi.org/10.1007/978-3-319-04468-2_10

West, J. (2014). *Natural disasters in Asia*. Asian Century Institute. Retrieved from http://asiancenturyinstitute.com/environment/40-natural-disasters-in-asia. Accessed 10 May 2017.

CHAPTER 2

SOCIAL CAPITAL AND CHANGES IN POST-DISASTER RECOVERY PROCESS: OBSERVATIONS FROM CHINA AFTER THE 2008 WENCHUAN EARTHQUAKE

Ziqiang Han

ABSTRACT

Disasters shape the development of communities and societies not only physically but also socially. This chapter provides some quantitative evidence to this effect by examining changes in social capital in communities affected by the 2008 Wenchuan Earthquake in China.

A two-wave, longitudinal household questionnaire survey data set was used for analysis. The baseline data were obtained in January, 2009, around 8 months after the 2008 Wenchuan Earthquake, adopting a stratified sampling method within a county severely disrupted by the earthquake. A follow-up survey with the same households was conducted in the summer of 2012. Finally, 415 household questionnaire surveys from nine communities within the county were collected for analysis.

Recovering from Catastrophic Disaster in Asia
Community, Environment and Disaster Risk Management, Volume 18, 15–36

ISSN: 2040-7262/doi:10.1108/S2040-726220160000018002

Overall, it can be concluded that social capital was strengthened in the post-disaster recovery process in the survey area. Social capital was measured according to three dimensions: (1) affiliation with organizations, (2) the degree of available social support, and (3) the degree of social cohesion within communities. It was found that the average degree of social capital increased during the recovery process, with a decrease of social capital inequalities between different families. More specifically, although informal personal networks were found to be the most prominent sources of social support, the support provided by formal organizations played a relatively more important role immediately after the catastrophe, given that most of the personal networks were also affected. Community cohesion was also found to have increased, with a decrease of standard deviation in the recovery process. This chapter suggests that disasters could generate positive effects rather than negative ones alone. Stronger and more tightly knit communities could be built in the disaster recovery process, if appropriate policies and methods are implemented.

Keywords: Social capital; disasters; social change; China; Wenchuan Earthquake

Social disruption is one of the most common and severe consequences of disasters besides life and economic losses. Although social changes after disasters have been a long-lasting theme in disaster research, there is a lack of consensus among the limited studies that exist. Some observations demonstrate that disasters could reshape communities, and others indicate that disasters would not change the pre-disaster social development trend. Researchers also argue that disasters can influence social change but that the direction and strength of this potential influence are dependent on the characteristics of the local or regional social structure (Lovekamp & Arlikatti, 2013). Current studies of social change following disasters contain two obvious limitations. Methodologically, most of these existing studies only use qualitative discourses without quantitative evidence. In addition, almost all of these studies have, to date, been conducted in Western societies, with a lack of evidence from non-Western societies. This chapter responds to these two limitations by examining the patterns of, and changes in, social capital after a natural disaster in China by using quantitative evidence—a longitudinal household survey data collected after the 2008 Wenchuan Earthquake.

This chapter first reviews existing literature on disaster and social change, as well as research on disasters and social capital. Then, the data collection procedures, measures, and analysis methods are presented. Following this, the results are presented, focusing on social capital changes in the disaster recovery process. The chapter ends with key conclusions and implications for future research.

DISASTER AND SOCIAL CHANGE

Disaster and social change is one of the most original research topics in the disaster research field. Disaster research pioneer, Samuel Prince's classic study that explored the reorganization process following the explosion of a munitions ship in the city of Halifax, first tackled this issue around a century ago. He found that the latter disaster created positive social change in terms of the city's aesthetic appearance, the expansion of business, population, political action, city planning, housing, health, education, recreation, and community spirit. It also generated a new sense of unity in terms of communities dealing with common problems after the disaster (Prince, 1920).

Views as to whether disasters can generate social change, however, are not consistent (Nigg & Tierney, 1993; Passerini, 2000). Macro-level changes after disasters, such as on the political, economic, and social levels, are arguably unlikely. When disasters do appear to effect social change, these changes tend to become manifest at a micro individual or small group level and are mostly negative, such as the increase of inequality within communities (Passerini, 2000). Nevertheless, this still points to the fact that changes always happen in the aftermath of disasters due to the direct impact of disasters, the disruptive consequences to society and communities, or the processes of readaptation (Carr, 1932). Some scholars have also confirmed that natural disasters do produce social change in terms of structure and complexity within communities (Hoover & Bates, 1985). Other researchers, however, have observed that the pre-disaster situations and development trends within communities have a more powerful influence than the disaster events themselves in the post-disaster recovery process. These arguments suggest that disasters only accelerate or decelerate the pre-disaster trends, without actually causing growing or prosperous communities to decline nor can they "save" declining communities (Nigg & Tierney, 1993). Despite the inconsistencies and ambiguities that have been found in terms of social change after disasters, both researchers (Nigg & Tierney, 1993; Passerini, 2000) who conducted reviews about disasters and social changes have admitted that there is no strong empirical evidence to support these claims.

"Social change" is, in itself, a vague term, which seems to refer to population changes, cultural changes, changes in the relationships between individuals or groups (Carr, 1932), or catastrophic changes. Harper and Leicht defined social change as "the significant alteration of social structure and cultural patterns through time" (2010, p. 5). Nigg and Tierney (1993) proposed five types of social change: (1) political/legal, (2) economic, (3) demographic/ecological, (4) social structural, and (5) quality of life changes, in their review of disasters and social changes; in this context, social structural changes included organizational changes and relationship changes.

It can be argued that mega disasters or catastrophes have more potential to create social changes. Maybe the soundest evidence is the 11 September 2001 terrorism attack in the United States. After the attack, the whole nation's emergency management efforts shifted to antiterrorism: the Department of Homeland Security, the Transportation Security Administration, the Homeland Security Advisory System, and the National Incident Management System were created, and the 2002 Homeland Security Act was passed (Lovekamp & Arlikatti, 2013).

This section mainly focuses on analyzing studies investigating social structural changes, especially relationship changes, in order to make the contents more related with this chapter's theme. Early scholars reported an increase of solidarity between family members or primary groups as a result of coping with disasters, a phenomenon described as the "therapeutic" (Fritz & Mathewson, 1957; Fritz & Rayner, 1958) or "altruistic" (Barton, 1969) effect of disasters. In a study conducted by Deutscher and New after the 1975 Kansas City tornado, the first ones to respond, such as fireman and police from neighboring communities who rushed to provide help, were not the only ones—even the sightseers who came for a "tornado adventure" experience applied their skills to the disaster relief. A feeling of "we-ness" and brotherhood emerged, and the forces of social cohesion resulting from the disaster situation appeared to transcend the immediate stricken area and the victims (Deutscher & New, 1961). In Drabek's study about disaster recovery following the 1966 tornado in Topeka, Kansas, the bonds between kin and friends were found to have become tighter, while linkages with neighbors were weakened (Drabek & Key, 1984). Recent observations after Hurricane Katrina in 2005 also found that prosocial behavior overwhelmingly emerged while antisocial behavior seemed relatively minor, in contrast to the imagery employed by the mainstream media (Rodriguez, Trainor, & Quarantelli, 2006).

Some researchers doubt that the prosocial behavior is always short lived, although they would emerge immediately after disasters (Passerini, 2000). Since natural disasters can create an external threat and community problems that need to be solved rapidly, and since the agent of the disaster can generally be perceived and specified, disasters can thus lead to a focusing of attention on the present and can, therefore, strengthen community cohesion and solidarity by providing extensive opportunities for participation in activities for the "good" of the community (Quarantelli & Dynes, 1976). In these situations, the presence of community conflict is rare, with very limited historical observations of the latter. If the allocation of blame, which is primarily found in manmade disasters, and the allocation of resources for rehabilitation are involved, there may be community conflict in the recovery period following a disaster (Quarantelli & Dynes, 1976). Moreover, even while prosocial behavior and a temporary cessation of pre-disaster conflict emerge, there may also be potential for conflicting changes within and between the different sectors of a community (Morrow & Peacock, 1997).

Recent studies have provided additional empirical evidence about social changes after disasters, especially from the perspective of social relationships and social capital. In the study that examined the survivors' quality-of-life recovery 5 years after the 2003 Bam Earthquake in Iran, a comparison of the quality of life between the disaster survivors and the general population revealed that the earthquake had resulted in better social relationships in the affected communities, although it may have adversely affected the psychological aspect of quality of life (Ardalan et al., 2011). In another study of social capital in communities affected by Hurricane Katrina in 2005, inequalities in social capital were found to have the potential to increase over the course of a disaster, from early preparation to prolonged displacement and uneven return. For example, poor communities were less likely to receive informal assistance, because their closest friends and kin network were in a similarly vulnerable situation (Elliott, Haney, & Sams-Abiodun, 2010). Evidence after the 2008 Wenchuan Earthquake in China also revealed a change of attitudes toward local government officials—these attitudes first declined because the promise of resource distribution from central government was not realized in local communities and then increased after the relief resources were delivered (Han, Hu, & Nigg, 2011).

Perhaps the most convincing survey comes from Dussaillant and Guzman's study about changes in the levels of trust in Chile after the 2010 earthquake. They used data constructed from several data sets collected during three different periods—a few months before the disaster, a few months after the disaster, and a year and a half later. The results showed that people's trust increased overall after the disaster, because new trust networks were promoted through the enhancement of distant family time and interaction between affected neighbors. Moreover, communities with higher levels of trust before the disaster were found to have a higher increased level after. This survey indicates that the trust-building effect of disasters can persist and increase over time, lasting almost 2 years after the disaster in the study (Dussaillant & Guzman, 2014). Another macro-analysis using cross-country panel data examined the relationship between social trust and disaster frequencies. The results indicate that natural disasters have the potential to encourage the emergence of a more tightly knit society (Toya & Skidmore, 2014).

In addition, disasters can enhance social capital elements such as trust and community cohesion through public participation in dealing with the same threat. However, such an effect is not consistent in academic observations because disasters may also create tense situations or conflict when resources are limited or their allocation is perceived as unfair. Most of these existing studies are qualitative in nature, with a lack of quantitative measures. In light of the latter, this chapter contributes to disaster and social change studies by referring to a quantitative dimension, namely, by analyzing survey results following the 2008 Wenchuan Earthquake in China.

DISASTER AND SOCIAL CAPITAL

The importance of social capital in disaster management and risk reduction has attracted the attention of disaster researchers in recent years (Dynes, 2006; Nakagawa & Shaw, 2004; Wisner, 2003). Compared with the devastation to physical and human capital typically enacted in disaster scenarios, social capital is seen to be less affected and can thus provide essential resources to accomplish critical tasks after disasters (Dynes, 2006). However, recent empirical studies indicate that the effects of social capital in this context could be complex and that social capital is actually a double-edged sword in disaster management (Aldrich, 2012a).

From one perspective, social capital plays an important role not only in the rescue and relief period in the immediate aftermath of a disaster but also in dealing with rehabilitation challenges (Nakagawa & Shaw, 2004). It has been found to be the best predictor of long-term post-disaster recovery (at least in terms of population growth), rather than the extent of earthquake damage, physical capital, human capital, or economic capital (Aldrich, 2012b). Formal and informal social networks have an important therapeutic and empowerment effect on helping disaster survivors, especially women, to overcome their post-disaster psychological traumas (Ganapati, 2012). They may also significantly affect individuals' utilization of assistance offered by formal organizations (Beggs, Haines, & Hurlbert, 1996) or informal sources (Zhao, 2007). Even within more diverse and loosely knit communities, social capital and community-based organizations can facilitate post-disaster recovery (Storr & Haeffele-Balch, 2012).

From another perspective, although social capital can facilitate decision making and collective actions (Aldrich, 2011; Storr & Haeffele-Balch, 2012), it may also block or delay urgent decision making (Buckland & Rahman, 1999). For example, after Hurricane Katrina in 2005, communities with a higher level of social capital (measured as voter turnout) resisted the set of unwanted public facilities (such as trailers) in their backyard and pushed "bad" public facilities into other communities (Aldrich & Crook, 2008). In addition, it has been found that inequalities in social capital can increase over the course of a disaster, reinforcing marginalized groups' obstacles to recovery (Aldrich, 2011) and making social disadvantages more pronounced. For example, people in the latter groups may not be able to evacuate on time due to a lack of private transit and inability to find proper accommodations; they also could be less likely to obtain informal assistance and return to their communities (Elliott et al., 2010).

For individuals or households, social capital can serve as "informal insurance" that provides information, financial help, and physical assistance (Aldrich, 2010). However, this effect may become less significant when individuals' social networks (i.e., bonding social capital) are widely impacted by catastrophes (Chowdhury, 2011; Fussell, 2006), or social capital itself is weakened by the marketization process in transforming societies (Lu, Sato, & Zhang, 2008).

On the basis of the preceding discussion, this chapter examines changes in household's social capital by using two-wave longitudinal survey data obtained after the 2008 Wenchuan Earthquake in China. By comparing the data collected in 2009 and 2012 from the same households, the author aims to examine the following hypotheses:

H1. Households' average social capital would increase in the recovery process.

H2. Inequalities in social capital would increase in the recovery process.

The current research also analyzes the correlations among financial capital, human capital, and social capital. In light of these, it is hypothesized that

H3. Social capital is positively related with a household's current financial status.

H4. Social capital is positively associated with a household's current human capital.

H5. Current social capital is more determined by the relative capital changes rather than the current status.

METHODS

Data Collection

The data used in this chapter were drawn from a longitudinal household survey, which was conducted in two data collection efforts. On 12 May 2008, a mega earthquake measuring 8.0 on the Richter scale struck the southwest of China, causing thousands of deaths and billions of Renminbi (RMB) worth of economic losses (National Center of Disaster Reduction, Ministry of Civil Affairs of China, & The United Nations Development Programme [NCDR & UNDP], 2009). In January, 2009, a household survey was conducted in W County, one of the most devastated counties according to the China National Disaster Reduction Council's assessment. The survey adopted a stratified sampling method, whereby all the towns within this county were divided into three groups – high, medium, and low – according to their economic condition. One town was randomly selected within each group. Then, three villages within each town were purposively selected according to their geographical locations, given that this region is a remote mountainous area and geographical location is thus significant to the overall condition of social development. One village from the mountaintop, one from the mountainside, and another from the valley area within each town were selected. Then, 60 households were surveyed within each village by using a convenience sampling method, with a final yield of 515 household surveys (60 households/village × 9 villages = 540 originally planned). A follow-up questionnaire survey with the same households was implemented in the summer of 2012, around 4 years after the disaster. Four hundred and fifteen households were successfully, albeit arduously, recontacted, yielding a follow-up rate of approximately 80%.

Both the questionnaire surveys were administrated by college students from the local area. About 15 college students who were local residents were hired each time, with the data collection efforts lasting around 15–20 days per questionnaire. When the students were hired, they were trained in understanding and applying the contents of the questions, communication and interview skills, and standards of recording the answers to the questions by the project's principle investigators. Finally, the collected questionnaire was transferred into a digital data set by using Epidata software and analyzed by using Stata 12.1.

Measurement

Social resources upon which people draw in pursuit of their livelihood objectives that developed through networks and connectedness, membership of more formalized groups, relationships of trust, reciprocity, and exchanges (Department for International Development [DFID], 1999) was adopted as the definition of social capital in this study. The World Bank's social capital initiative efforts (Grootaert, 2004; Grootaert & Van Bastelaer, 2002) in the early 21st century developed six dimensions of social capital: groups and networks, trust and solidarity, collective action and cooperation, information and communication, social cohesion and inclusion, and empowerment and political action. On the basis of these measurement efforts and the realities in the field, the current study adopted three groups of proxy indicators as measurements of social capital: organizational embeddedness, social support, and community cohesion.

Organizational embeddedness itself contains three variables: organizational affiliation (number of organizational membership), degree of activity (whether actively involved in the organization), and leadership status (whether or not the respondent holds a leading position in the organization). Both the social support and social cohesion indicators were developed on the basis of the World Bank's aforementioned social capital initiative efforts, as well as from similar research efforts in China (Qiu, Chen & Ren, 1998; Sun, Xu, Tao & Su, 2007). The social support indicator captured the social support available to local people from nine proposed sources, which included both informal social networks such as friends, neighbors, close relatives, or other relatives and formal organizations such as village committees, the town government, banks, local financial organizations, or other organizations. The social cohesion and trust indicator was measured by the participant's response to six statements. Detailed measurements are illustrated in the following section, alongside the presentation of the results.

RESULTS: SOCIAL CAPITAL PATTERNS AND CHANGES

This section presents key patterns and changes in the selected indicators of social capital—organizational embeddedness, social support, and social cohesion within the local community. This is followed by a comparison of the social capital indicators between 2009 and 2012. Finally, the key factors found to have influenced social capital are explored and discussed.

Organizational Embeddedness

Three closely related variables were used to construct the organizational embeddedness indicator. The first one was a dummy variable representing whether or not a family member held organizational membership (coded as 1 or 0) from a list of 12 proposed formal organizations[1] based on prior field work and studies. The second variable measured respondents' activity (active or not, coded as 2 or 1) within the organization, and the third one captured the respondent's leadership status (whether or not they hold a leading position, coded as 2 or 1) with the organization. Then, the product of the three variables was calculated as each family member's organizational embeddedness score. Finally, the sum of each family member's organization embeddedness score was used as the organization embeddedness indicator of each family as a whole.

The level of organizational embeddedness before the earthquake was also collected through respondents' retrospective views obtained during our first questionnaire survey. Thus, the final data obtained actually revealed the status of organizational embeddedness at three time periods: before the earthquake, soon after the earthquake (8 months later), and 4 years after the earthquake.

The frequency distribution and the proportion of organizational embeddedness found are reported in Table 1. Overall, respondents' participation in formal organizations was found to display an increasing trend since the disaster, with more families reporting organizational membership since the earthquake. Leadership and activity within these organizations decreased slightly after the disaster; however, this trend stopped and recovered in the long run. Before the earthquake, 31.08% of the 415 families in our sample had family members in organizations. This number increased to 38.07% after the earthquake at the beginning of 2009, rising further to 48.92% in 2012. Among families who had members involved in formal organizations, most of them only had one family member involved, with very few reporting three or more. The changing patterns among these three types of families (family with one member, family with two members, or family with three or more members in organizations) were similar to the overall change pattern of the number of families that had an organizational affiliation, both of which maintained a stable increase following the earthquake. Of all the families with some kind of organizational membership, 21 had family members who had held a leadership position in the organizations before the earthquake, 18 had family members with a leadership position 8 months after the disaster, and 20 had family members with a leadership position about 4 years later. Of the total 415 families followed up, 29.88% of them reported that they had at least one

Table 1. Organizational Embeddedness.

	Pre-earthquake		Year 2009		Year 2012	
	Freq.	Percent	Freq.	Percent	Freq.	Percent
Number of family members in organizations						
None	286	68.92	257	61.93	212	51.08
One	81	19.52	86	20.72	102	24.58
Two	35	8.43	47	11.33	57	13.73
Three or more	13	3.13	25	6.02	44	10.6
Leading position held in organization/s						
None	394	94.94	397	95.66	395	95.18
One	21	5.06	18	4.34	20	4.82
Number of active members						
None	291	70.12	298	71.81	273	65.78
One	77	18.55	72	17.35	101	24.34
Two	34	8.19	32	7.71	30	7.23
Three or more	13	3.13	13	3.13	11	2.65
Total	415	100	415	100	415	100

family member active in organizations. This percentage changed to 28.19% in January 2009 but increased even further to 34.22% in 2012.

A detailed investigation of our data indicates that the development of formal organizations has been limited in rural China. In this sample, only four kinds of local organizations were reported among the respondents, even though we initially proposed 12 possibilities. Two of the reported organizations were political (the Communist Party and the Communist Youth League local chapters), and the other two were faith based (Buddhism and Christianity).

Social Support

The question "If your family were in difficulty, could you get help from the following sources?" was used to capture respondents' degree of social support. Nine possible sources were proposed. The five-item Likert scale was used, with values from one to five indicating "definitely cannot get help," "maybe cannot get help," "not sure," "may be able to get help," and "definitely can get help," respectively. The average response rates for each of these questions are reported in Table 2. The principal-component factor analysis

Table 2. Scores by Sources of Social Support.

Social Support	Year 2009		Year 2012	
	Mean	Std. Dev.	Mean	Std. Dev.
Friends	3.84	1.15	3.87	1.32
Neighbors	3.83	1.09	3.94	1.28
Close relatives such as children/parents	2.60	2.34	3.34	2.18
Other relatives	3.88	1.08	4.27	1.08
Village committee	2.80	1.24	2.73	1.40
Town government	2.54	1.16	2.11	1.15
Banks	2.27	1.16	2.17	1.23
Local financial organizations	2.31	1.19	2.04	1.08
Others	1.55	1.43	2.24	1.08

Question: "If your family was in difficulty, would you be able to get help from the following sources?"
Answers: 1, "definitely cannot get help"; 2, "maybe cannot get help"; 3, "not sure"; 4, "may be able to get help"; 5, "definitely can get help."

method was then used to explore the clustered attributes of these nine items. The factor loadings and unique variances following an orthogonal varimax (Kaiser off) rotation are presented in Table 3.

Overall, it can be seen that respondents obtained more social support from their personal social networks such as friends, neighbors, and relatives than from formal organizations such as local government and financial organizations, not only in times of crisis but also during normal times. In both survey waves, the mean values of support scores from personal social networks were above 3.0 (with the exception of the close relatives in 2009), whereas the mean values of support scores from formal organizations were below 3.0.

Although informal personal networks were found to provide more social support than formal organizations in general, their relative importance in terms of support provision during various stages of disaster recovery varied. Measured by the mean values of the support score obtained for each source, people reported that they had more support from informal networks in 2012 than in 2009 and more support from formal organizations in 2009 than in 2012. This result implies that formal organizations provide more support to disaster survivors during crisis time than normal time, most likely because most survivors' personal networks were themselves impacted by the catastrophic earthquake.

Table 3. Factor Loadings and Unique Variances of Social Support Sources.

Variable	Year 2009				Year 2012		
	Factor1	Factor2	Factor3	Uniqueness	Factor1	Factor2	Uniqueness
Friends	**0.84**	0.11	0.08	0.27	**0.84**	0.05	0.30
Neighbors	**0.85**	0.18	0.04	0.24	**0.87**	0.03	0.24
Close relatives such as children/parents	0.19	0.12	**0.67**	0.51	0.40	−0.08	0.84
Other relatives	**0.73**	0.02	0.16	0.44	**0.79**	−0.03	0.37
Village committee	0.41	**0.58**	−0.02	0.50	0.48	0.39	0.62
Town government	0.28	**0.75**	0.03	0.36	0.13	**0.69**	0.51
Banks	0.04	**0.90**	0.13	0.18	0.00	**0.89**	0.21
Local financial organizations	0.05	**0.87**	0.08	0.23	−0.01	**0.90**	0.20
Others	0.06	0.12	**0.81**	0.32	0.02	0.47	0.78

Method: Principal-component factors analysis.
Rotation: Orthogonal varimax (Kaiser off).
Cumulative proportion: 0.6604 (Year 2009); 0.5476 (Year 2012).

As can be seen from the results, the standard deviation changes of all the social support scores were not consistent. The standard deviations relating to friends, neighbors, village committees, and banks increased from 2009 and 2012, whereas the same indicator with regard to relatives, town government, and local financial organizations decreased.

The factor analysis results revealed certain differences and patterns in the social support provided between the beginning of recovery in 2009 and 3 years later in 2012. For the 2012 data, two main component factors explaining 54.76% of the variance emerged. However, for the 2009 data, three factors were retained after the principal-component factor analysis, the cumulative explanative power of which was 0.6604. Friends, neighbors, and other relatives were clustered as the informal support factor, while town government, banks, and other local financial organizations were grouped as the formal support sources in both waves of data. Village committees swung between the informal factor and formal factor. During the early recovery period, support from village committees loaded more on the formal support, but during normal times in 2012, it was perceived more as an informal support source. Unlike the situation during normal times, when close relatives such as children and parents hold similar weight to other informal supports, close relatives came out as a separate factor with others (e.g., volunteers) in the 2009 data, 8 months after the catastrophe. Most of the "others" were reported as volunteers here. This result indicates that during crises or emergencies, close relatives or family members could provide special social support to disaster survivors and that this support is different from other sources of informal support or formal assistance. In addition, it can be argued that survivors always hold gratitude for volunteers, especially after they have experienced such tragedy. In our study, volunteers were perceived as close as intimate relatives, not only because "other" and "close relatives" were grouped together but also because they had a similar mean value in 2009 (the average value of the "others" would be 2.64 if the "not applicable" data were to be excluded).

Social Cohesion

The social cohesion within communities was measured by respondents' rates of agreement with six statements (shown in Table 4). Similar to the measurement of social support, the answers to these six statements used a Likert scale ranging from one to five, representing "totally disagree" to "totally agree," respectively. As shown in Table 4, all the mean values of the six statement responses increased from 2009 to 2012, suggesting that the social cohesion within communities improved during the disaster recovery process from the

Table 4. Social Cohesion Within Community.

Variable	Year 2009		Year 2012	
	Mean	Std. Dev.	Mean	Std. Dev.
1. If you needed help, most of the people in this village would help you	3.79	0.82	4.09	0.80
2. If you sacrificed for the public good, other villagers would support you	3.57	0.88	4.09	0.86
3. Most of the people in this village can be trusted	3.69	0.70	3.83	0.66
4. If someone was in need of money for an emergency, most people would lend to him/her	3.54	0.78	3.65	0.75
5. If you had children, you would ask your neighbors to take care of them when you were not available	3.53	0.90	4.00	0.75
6. If your family had a wedding or a funeral, most of the villagers would come to help	4.25	0.69	4.61	0.60

Question: To what extent do you agree or disagree with the following statements?
Answers: 1, Totally disagree; 2, Disagree; 3, No opinion; 4, Agree; 5, Totally agree.

Table 5. Social Cohesion: Factor Loadings and Uniqueness.

Variable	Year 2009		Year 2012		
	Factor	Uniqueness	Factor1	Factor2	Uniqueness
1. If you needed help, most of the people in this village would help you	0.76	0.43	0.73	0.24	0.41
2. If you sacrificed for the public goods, other villagers would support you	0.69	0.53	0.85	−0.04	0.27
3. Most of the people in this village can be trusted	0.78	0.39	0.45	0.49	0.56
4. If someone was in need of money for an emergency, most people would lend to him/her	0.69	0.52	0.00	0.93	0.14
5. If you had children, you would ask your neighbors to take care of them when you were not available	0.52	0.73	0.60	0.16	0.61
6. If your family had a wedding or a funeral, most of the villagers would come to help	0.55	0.70	0.68	0.01	0.54

Method: Principal-component factors analysis.
Rotation: Orthogonal varimax (Kaiser off).
Cumulative proportion: 0.4504 (Year 2009); 0.5781 (Year 2012).

beginning of 2009 to the summer of 2012. Meanwhile, all the standard deviations in 2012 were found to be smaller than the standard deviations in 2009, indicating that people's consent with these statements was more homogeneous in 2012 than in the early recovery period.

To explore the patterns of social cohesion within communities, we applied principal-component factors analysis. The factor loadings and uniqueness values following an orthogonal varimax rotation are reported in Table 5. Only one factor was generated from the first data set in 2009, but two factors were retained in the data from 2012. The finance-related statement was separated from the others. Furthermore, the trust-related statements had similar loadings in both factors in the analysis using the 2012 data.

These changing patterns arguably reflect changes in people's values, especially their attitudes toward money after this catastrophe. In a more secular, materialistic, and underdeveloped society, such as rural areas in the west of China, people value money more than their time or labor simply because money is a relatively more scarce resource. For example, it is not difficult to ask villagers to contribute their labor or time for public affairs, but it is very difficult to collect cash for public goods. Moreover, although it is well known that the

Chinese have a culture of saving money, this attitude toward money changed for a while among the disaster survivors after the tragedy. People valued their quality of life and time with family more than ever. For example, as some people said, "Life is unpredictable, what is the meaning of savings for us if we died in the earthquake?" One interesting story we heard from several people during our first and second visits to the disaster-impacted area was that the sale of luxury cars increased sharply after the earthquake. A significant number of rich people bought new cars and went traveling in the wake of the catastrophe. These changes in people's attitudes are reflected in our data. When life went back to normal in 2012, people started thinking about their future and began to treat financial issues more seriously, especially after they spent almost all their savings on their new houses or became indebted due to reconstruction.

$$\text{Social capital index} = \sum (\text{Organization participation (leadership} \times \text{activity} \times \text{membership), Mean (social support score), Mean (community cohesion score))} \quad (2.1)$$

Social Capital Changes and Influencing Factors

The social capital indicator was generated by the aggregation of the organizational embeddedness score, the average social support score, and the average community cohesion score presented in Equation (2.1). The organizational embeddedness indicator of a family was taken as the sum of all of the family members' organizational participation score. For each family member, the organizational participation score was the product of their leadership (whether they held a leading position in the organization, 2 or 1), their degree of activity in the organization (active or not, 2 or 1), their actual membership (0 or 1) within an organization, and the number of organizations they participated in. The social support indicator was taken as the average of the scores obtained for the nine social support sources presented in Table 2. Similarly, the mean of the six social cohesion scores from Table 4 was used as the community cohesion indicator.

Overall, it was found that social capital had increased during the disaster recovery process. The average social capital indicator increased by 9.27%, while the standard deviation of the social capital indicator actually decreased, indicating that inequalities in social capital between the earthquake survivors had narrowed from the very beginning of the recovery period in 2009 to the summer of 2012. Meanwhile, the three social capital subindicators displayed a similar trend to the overall social capital indicator, with the average values increasing while the gaps decreased (Table 6).

Table 6. Social Capital Indicator and Sub-indicators.

Variable	Mean			Std. Dev.		
	2009	2012	Change Rate	2009	2012	Change Rate
Social capital indicator	7.796	8.519	9.27%	2.292	2.291	−0.04%
Organization indicator	1.188	1.506	26.77%	2.013	2.012	−0.05%
Social support indicator	2.879	2.968	3.09%	0.744	0.704	−5.38%
Cohesion indicator	3.729	4.045	8.47%	0.529	0.465	−12.10%

Table 7. Social Capital and Influencing Factors.

Variable	Social 2009	Social 2012A	Social 2012B	Social_dd
Income2009	−0.000		−0.000	
Human2009	0.723***		0.087	
Land2009	0.020		0.015	
House2009	0.000		0.000	
Income2012		0.000**	0.000**	
Human2012		0.601***	0.316**	
Land2012		0.008	0.003	
House2012		0.000	0.000	
Social2009			0.398***	
Income_dd				0.000
Human_dd				0.079
Land_dd				−0.001
House_dd				0.000

Notes: * $P < 0.05$; ** $P < 0.01$; *** $P < 0.001$.

Different ordinary least squares (OLS) regression models (Table 7), which adopted the constructed social capital indicator as the dependent variable, were used to explore the factors influencing social capital. Household annual income, family levels of human capital, land owned by families, and the estimated value of houses that families owned were used as predicting variables. Since the data set is composed of two-wave panel data, both the social capital indicators in 2009 and 2012, respectively, and their difference were used as dependent variables. For the first model (Social 2009), the social capital indicator in 2009 was used as the dependent variable, and the income, human capital, land owned by family, and estimated house value in 2009 were used as predicting variables. Similarly, the second model (Social 2012A) used data from 2012. The third model (Social 2012B) used the social capital indicator in 2012 as the dependent variable but included all other variables from both 2009 and 2012 as predicting variables. The last model (Social_dd) adopted the difference-in-difference method, which used the differences between the 2012 and 2009 data as both dependent and independent variables.

The results from these models indicated that the current human capital status in the affected region and the prior social capital indicator are the most significant predictors of current social capital. The changes in the effects of these independent variables were not significant. In other words, families' social capital can be said to be more influenced by their current human capital status and prior social capital, rather than the changes that occurred from 2009 to 2012 (i.e., the difference between the values of 2009 and 2012). The financial variable was found to have a very weak positive effect on the social capital indicator, which effect is only statistically significant in the 2012 data.

DISCUSSION AND CONCLUSION

Following debates about disasters and social changes in Western cultures (Nigg & Tierney, 1993; Passerini, 2000), the current analysis demonstrates that there does, indeed, exist some kind of social change in the post-disaster recovery process from the social capital perspective. Specifically, in our study, the average value of social capital held by household was found to have increased (Hypothesis H1), while inequalities in social capital between different households were found to have decreased (Hypothesis H2) in the recovery process of the area affected by the 2008 Wenchuan Earthquake from 2009 to 2012.

Disasters have the ability to foster micro-level, household, and community changes (Lovekamp & Arlikatti, 2013). The three indicators of social capital used in the current research – organizational embeddedness, social support, and community cohesion – were all found to have similar change trends to those of changes in overall social capital. Informal social networks of individuals were found to play a more important role in their social support than formal organizations; however, the patterns of social support examined reveal that support of formal organizations was more important during the time of crisis itself than normal times, because most social networks of individuals are themselves locally embedded and were thus impacted during the catastrophe. Social cohesion within the communities studied was found to have become stronger during the recovery process. Also, the clustered patterns of social capital measures have a little change. During emergencies or immediately after disasters, people treated finance, trust and labor, among other variables, more similarly than in normal times. With the recovery process eventually reverting the communities back to their normal status, the financial and trust-related culture of cohesion was separated out subconsciously from other social cohesion-influencing factors.

Many prior studies have confirmed that social capital is a key influencing factor of recovery outcomes after disasters (Aldrich, 2012a; Marín, Bodin, Gelcich, & Crona, 2015; Storr, & Haeffele-Balch, 2012) and can also empower vulnerable groups within communities (Lovekamp & Arlikatti, 2013). In addition to using social capital as a predicting variable, this chapter has explored the correlation between social capital and households' human capital, as well as financial status. One aim was to ascertain which types of disaster victims have higher degrees of social capital. The regression results demonstrate that human capital and financial status are the two positive predictors of the current social capital held by affected families (H3 and H4). Current human capital was found to be the most consistent positive predictor of social capital (H5)—an effect that was significant in both models using the 2009 data and the 2012 data separately. Current levels of social capital were also seen to be positively influenced by families' current financial status, including factors such as annual income (H4), but this effect was only significant in the model using the 2012 data. Moreover, the models' results indicate that current levels of social capital are more determined by current human capital and financial status, rather than relative changes in capital (H3).

This chapter has explored patterns and changes in social capital by using data from longitudinal household surveys conducted in rural China after the 2008 Wenchuan Earthquake. In conclusion, the changes in social capital found from 2009 to 2012 (higher means, lower standard deviations) indicate that the social atmosphere becomes better in the post-disaster recovery process. The regression results indicate that families with higher degrees of human capital and a higher financial status would have higher social capital. With these conclusions, this chapter has contributed to the current knowledge on disaster and social changes from a social capital perspective and, moreover, from a non-Western culture viewpoint. In terms of future research in this field, studies with a more rigorous design in terms of measurement and sample size could be implemented.

NOTE

1. The 12 proposed organizations were the Communist Party (local branches), the Communist Youth League (local branches), community watch guards, the Women's Association, the Workers Union, Village Autonomous Committee, community voluntary organizations, the Elders Association, religious organizations, recreation or hobby-based organizations, agriculture producer cooperatives, and other NGOs/NPOs.

ACKNOWLEDGMENT

This study was funded by Sichuan University under Grant YJ201505.

REFERENCES

Aldrich, D. (2012a). *Building resilience: Social capital in post-disaster recovery*. Chicago, IL: University of Chicago Press.

Aldrich, D. (2012b). Social, not physical, infrastructure: The critical role of civil society after the 1923 Tokyo earthquake. *Disasters*, *36*(3), 398–419.

Aldrich, D. P. (2010). Fixing recovery: Social capital in post-crisis resilience. *Journal of Homeland Security*, *6*, 1–10.

Aldrich, D. P. (2011). The externalities of strong social capital: Post-tsunami recovery in southeast India. *Journal of Civil Society*, *7*(1), 81–99.

Aldrich, D. P., & Crook, K. (2008). Strong civil society as a double-edged sword: Siting trailers in post-Katrina New Orleans. *Political Research Quarterly*, *61*(3), 379–389.

Ardalan, A., Mazaheri, M., Vanrooyen, M., Mowafi, H., Nedjat, S., Naieni, K. H., ... Russel, M. (2011). Post-disaster quality of life among older survivors five years after the Bam earthquake: Implications for recovery policy. *Ageing & Society*, *31*(Part 2), 179–196.

Barton, A. H. (1969). *Communities in disaster: A sociological analysis of collective stress situations*. Garden City, NY: Doubleday.

Beggs, J. J., Haines, V., & Hurlbert, J. S. (1996). The effects of personal network and local community contexts on the receipt of formal aid during disaster recovery. *International Journal of Mass Emergencies and Disasters*, *14*(1), 57–78.

Buckland, J., & Rahman, M. (1999). Community-based disaster management during the 1997 Red River Flood in Canada. *Disasters*, *23*(2), 174–191.

Carr, L. J. (1932). Disaster and the sequence-pattern concept of social change. *American Journal of Sociology*, *38*(2), 207–218.

Chowdhury, M. J. A. (2011). *Social cohesion and natural disaster loss recovery of households: Experience from Bangladesh*. Working Paper, University of Dhaka, Bangladesh.

Department for International Development [DFID]. (1999). *Sustainable livelihoods guidance sheets*. London: DFID.

Deutscher, I., & New, P. K. (1961). A functional analysis of collective behavior in a disaster. *The Sociological Quarterly*, *2*(1), 21–36.

Drabek, T. E., & Key, W. H. (1984). *Conquering disaster: Family recovery and long-term consequences*. New York: Irvington Publishers.

Dussaillant, F., & Guzman, E. (2014). Trust via disasters: The case of Chile's 2010 earthquake. *Disasters*, *38*(4), 808–832.

Dynes, R. R. (2006). Social capital: Dealing with community emergencies. *Homeland Security Affairs*, *2*(2), 1–26.

Elliott, J. R., Haney, T. J., & Sams-Abiodun, P. (2010). LIMITS TO SOCIAL CAPITAL: Comparing network assistance in two New Orleans neighborhoods devastated by Hurricane Katrina. *Sociological Quarterly*, *51*(4), 624–648.

Fritz, C. E., & Mathewson, J. H. (1957). *Convergence behavior in disasters: A problem in social control* (a special report prepared for the Committee on Disaster Studies). National Academy of Sciences National Research Council.
Fritz, C. E., & Rayner, J. F. (1958). *The therapeutic features of disaster and their effect on family adjustment: some research orientations*. Paper presented at the Conference on Marriage and the Family, Washington, DC.
Fussell, E. (2006). Leaving New Orleans: Social stratification, networks, and hurricane evacuation. Understanding Katrina: Perspectives from the Social Sciences, the Forum of the Social Science Research Council. Retrieved July 10, 2016 from http://understandingkatrina.ssrc.org/Fussell/
Ganapati, N. E. (2012). In good company: Why social capital matters for women during disaster recovery. *Public Administration Review*, *72*(3), 419–427.
Grootaert, C. (2004). *Measuring social capital: An integrated questionnaire*. Washington, DC: World Bank Publications.
Grootaert, C., & Van Bastelaer, T. (2002). *Understanding and measuring social capital: A multidisciplinary tool for practitioners*. Washington, DC: World Bank Publications.
Han, Z., Hu, X., & Nigg, J. (2011). How does disaster relief works affect the trust in local government? A study of the Wenchuan earthquake. *Risk, Hazards & Crisis in Public Policy*, *2*(4), 1–20.
Harper, C. L., & Leicht, K. T. (2010). *Exploring social change: America and the world* (6th ed.). Boston, MA: Routledge.
Hoover, G. A., & Bates, F. L. (1985). The impact of a natural disaster on the division of labor in twelve Guatemalan communities: A study of social change in a developing country. *International journal of mass emergencies and disasters*, *3*(3), 7–26.
Lovekamp, W. E., & Arlikatti, S. (2013). Social change and empowerment. In D. S. K. Thomas, B. D. Phillips, W. E. Lovekamp, & A. Fothergill (Eds.), *Social vulnerability to disasters* (2nd ed., pp. 447–473). Boca Raton, FL: CRC Press (Reprinted).
Lu, M., Sato, H., & Zhang, S. (2008). Does social capital help households cope with natural disaster during marketization? Evidence from rural China. Paper presented at the *Emergence of China and India in the Global Economy Conference*, July 3–5, 2008, London, UK.
Marín, A., Bodin, Ö., Gelcich, S., & Crona, B. (2015). Social capital in post-disaster recovery trajectories: Insights from a longitudinal study of tsunami-impacted small-scale fisher organizations in Chile. *Global Environmental Change*, *35*, 450–462. Retrieved from https://doi.org/10.1016/j.gloenvcha.2015.09.020
Morrow, B. H., & Peacock, W. G. (1997). Disasters and social change: Hurricane Andrew and the reshaping of Miami. In W. G. Peacock, B. H. Morrow, & H. Gladwin (Eds.), *Hurricane Andrew: Ethnicity, gender and the sociology of disasters* (pp. 226–242). New York: Routledge.
Nakagawa, Y., & Shaw, R. (2004). Social capital: A missing link to disaster recovery. *International Journal of Mass Emergencies and Disasters*, *22*(1), 5–34.
National Center of Disaster Reduction, Ministry of Civil Affairs of China, & The United Nations Development Programme (NCDR & UNDP) (2009). Early Recovery and Disaster Risk Management Program. Research Report on Disaster Rescue and Relief in Wenchuan Earthquake.
Nigg, J. M., & Tierney, K. J. (1993). Disasters and social change: Consequences for community construct and affect. Paper presented at the *Annual Meeting of the American Sociological Association*, Miami, Florida.

Passerini, E. (2000). Disasters as agents of social change in recovery and reconstruction. *Natural Hazards Review*, *1*(2), 67–72.

Prince, S. H. (1920). *Catastrophe and social change: Based upon a sociological study of the Halifax Disaster* (Vol. 94). New York, NY:Columbia University.

Qiu, H., Chen, J., & Ren, Y. (1998). Social support structures and changes: From one dimension to multi-dimensions. *Sociological Studies*, *04*, 33–39 (丘海雄, 陈健民, & 任焰. (1998). 社会支持结构的转变: 从一元到多元. *社会学研究*(*04*), 33–39).

Quarantelli, E. L., & Dynes, R. R. (1976). Community conflict: Its absence and its presence in natural disasters. *Mass Emergencies*, *1*, 139–152.

Rodriguez, H., Trainor, J., & Quarantelli, E. L. (2006). Rising to the challenges of a catastrophe: The emergent and prosocial behavior following Hurricane Katrina. *The Annals of the American Academy of Political and Social Science*, *604*(1), 82–101.

Storr, V. H., & Haeffele-Balch, S. (2012). Post-disaster community recovery in heterogeneous, loosely connected communities. *Review of Social Economy*, *70*(3), 295–314.

Sun, X., Xu, Z., Tao, R., & Su, F.(2007). Political trust, social capital and participation of rural election—An analysis of a national data. *Sociological Studies, 04*, 165–187. (孙昕, 徐志刚, 陶然, & 苏福兵. (2007). 政治信任、社会资本和村民选举参与——基于全国代表性样本调查的实证分析. *社会学研究*(*04*), 165–187.)

Toya, H., & Skidmore, M. (2014). Do natural disasters enhance societal trust? *KYKLOS*, *67*(2), 255–279.

Wisner, B. (2003). Disaster risk reduction in megacities: making the most of human and social capital. In A. Kreimer, M. Arnold, & A. Carlin (Eds.), *Building safer cities: The future of disaster risk* (pp. 181–196). Washington, DC: World Bank.

Zhao, Y. (2007). Social capital and post-disaster recovery: A sociological study of natural disaster [J]. *Sociological Studies*, *05*, 164–187.

CHAPTER 3

RECOVERING FROM PROLONGED NEGATIVE DESTINATION IMAGES IN POST-DISASTER NORTHERN JAPAN

David N. Nguyen and Fumihiko Imamura

ABSTRACT

The Tohoku Region in northern Japan was devastated by the 2011 Great East Japan Earthquake and Tsunami. Although much of the physical infrastructure has since been restored, annual tourism numbers have yet to restore to pre-disaster levels. In some areas, tourism recovery remains stagnant. The objective of this research is to examine how local decision-makers utilize media strategies to deal with image-related crises and reverse negative images to combat stereotypes and deliver successful campaign messages. The study has found that each of the prefectures affected by the disasters has since utilized different campaign strategies with some, such as Fukushima, focusing on the future, while others, such as Aomori, utilized a mixture disassociating itself from the troubled area and associating itself with its more prestigious neighbor. Much of these negative images stem from persisting images of region-wide safety fears over natural hazards and radiation concerns. This study suggests that further research needs to be

Recovering from Catastrophic Disaster in Asia
Community, Environment and Disaster Risk Management, Volume 18, 37–59

ISSN: 2040-7262/doi:10.1108/S2040-726220160000018003

done to identify the different risk perceptions of foreign tourists by country, as some groups such as Koreans are more risk averse leading to a sharp decline in visits, while others such as Taiwanese who are accustomed to natural hazards are leading in visitor numbers and tourism recovery.

Keywords: Place-marketing; negative image; tourism recovery; disasters; Japan

INTRODUCTION

Tourism is a major source of income, investment, and employment for coastal communities. The World Trade Organization (WTO) believes that tourism has the capacity to stimulate development, economic growth, and create new opportunities for poverty alleviation and self-governance (WTO, 2010). Within the Asia-Pacific, international tourism has grown significantly where much of the tourism is intraregional and domestic.

However, economic and tourism developments exhibit strong core–periphery relationships where tourism may play a role to facilitate the development of these peripheral areas. These relationships not only exist between countries but also within a subnational space (Wall, 2008). Tourism is utilized as a tool for the accumulation of capital in peripheral regions with rich tourists traveling from the metropolitan center to the periphery, bringing foreign exchange and creating jobs. The state, in particular, plays an important role in regional tourism development as it is often responsible for the proposals for regional development schemes (Telfer, 2002). As such, tourism has been used to revitalize areas in economic decline, as well as distant territories such as islands, through the use of government investments and incentives toward tourism businesses.

The tourism economy, however, contains a number of vulnerabilities that are unique to the industry. As destinations rely on the intrinsic assets of the local environment, tourism sites are often located near areas that are exposed to natural hazards such as high winds, volcanic activity, storm surges, tsunamis, and sea-level rise. In addition, as tourism is a major user of local infrastructure, which includes transportation networks, electrical systems, and water-supply systems, disruption to these services can have negative short-term and long-term repercussions that may lead to eroding destination image (Byrd, 2007; Huan, Beaman, & Shelby, 2004).

An example of the natural hazard vulnerabilities in tourism can be seen after the 2004 Indian Ocean Tsunamis that directly devastated six countries in the Indian Ocean. In the case of Thailand, the tsunami caused the country's worst natural disaster with over 5,300 confirmed deaths, nearly 3,000 missing, and losses accounting nearly a quarter of the national gross domestic product (GDP) (Rosa, 2012). Heavily dependent on the tourism industry, the tsunami was a major blow to the Thai economy as half of the confirmed deaths consisted of tourists (Chanchai, Chao, & Omthip, 2006). Although tourists have maintained positive impressions of Thailand, Phuket's association with the tsunami could lead to tourists traveling to Pattaya as an alternative location that offers similar amenities while being perceived as a safer destination (Tavitiyaman & Qu, 2013).

Communities rely on continued visitor numbers for their livelihoods, and delayed tourism recovery can cause severe impacts beyond economic effects (Mair, Ritchie, & Walters, 2016). Thus, it is imperative for destinations to respond promptly and effectively in the wake of a disastrous event to limit the impacts on the local tourism industry. As tourism relies on the promotion of a safe and stable environment, despite its vulnerabilities to natural hazards and other external shocks (Nguyen, Imamura, & Luchi, 2016), how should destinations approach the recovery of its visitor numbers and reputation while images of destruction and vulnerabilities continue to proliferate in the mass media?

This study examines the initiatives and messaging strategies utilized by prefectural tourism agencies in the Tohoku Region of northeastern Japan, which was devastated by the 2011 Great East Japan Earthquake and Tsunami. Before the disasters, the region was already reeling from negative images associated with its peripheral location and prolonged economic decline. In addition, Tohoku continues to reel under radiation fears caused by the meltdown of the Fukushima Daiichi Nuclear Power Plant. Tourism planners face immense challenges in convincing tourists to return to the region as they not only have to deal with recovering from short- to medium-term negative images but also long-term ones that are more difficult to address.

LITERATURE REVIEW

Although the media uses the term "disasters" and "crises" interchangeably, Faulkner distinguishes between the two, the former is referring to a situation where a destination is confronted with a sudden unpredictable catastrophic event to which it has little control (Faulkner, 2001). Lew refers to these events as one of the many external shocks, which were categorized into five levels, ranging from those that are frequent and expected such as currency fluctuations to those that are highly unlikely but can catch the tourism stakeholders off guard, such as large-scale natural disasters (Lew, 2013). On the other hand, crisis is referred to a situation where the root cause of an event stems primarily from self-inflicted problems such as poor management structures, failure to adapt to change, and inept planning (Faulkner, 2001).

Tourism recovery thus focuses on marketing to restore tourism confidence to a destination with a goal to return to pre-disaster level (Lehto, Douglas, & Park, 2008). Literature identifies that marketing communication must focus on two areas to restore visitor numbers, changing the perceptions and misconceptions of the destination caused by intense media coverage, and restoring visitor confidence in the destination (Prideaux, Coghlan, & Falco-Mammone, 2008; Scott, Laws, & Prideaux, 2008; Walters & Mair, 2012). Walters and Mair state that the former appears to change attitudes while the latter may focus more on strengthening an existing positive attitude. However, there is little agreement on how to best achieve these goals (Walters & Mair, 2012). Literature also suggests that the possibility that once damage has been done, there may be little left to attract visitors (Cioccio & Michael, 2007).

Avraham and Ketter identify two types of negative destination images. The first are ones caused by an unexpected crisis such as terrorism, natural disasters, or sudden epidemics. Second are those that are prolonged caused by long-lasting problems such as economic hardships, high crime rates, or continuous instability. Combating these images can be challenging due to years of enduring stereotypes and prejudices that are also reinforced in the mind by superficial and sensational media coverage (Avraham & Ketter, 2013). It is unrealistic to expect any target audience to visit a destination and ignore these long-lasting negative images and stereotypes just because of an ad campaign or other promotional tools (Anholt, 2006). These stereotypes represent simplified attitudes or beliefs about a place that are not examined thoroughly and are difficult to change. Once labeled, it begins a symbol of these activities, and other events in this place are less likely to receive any media coverage (Shields, 1992). Thus far, existing literature on marketing solutions has tended to focus on short- to mid-term crisis rather than prolonged ones (Avraham & Ketter, 2008b).

To mitigate these negative images, Avraham and Ketter proposed the "Multi-Step Model for Altering Place Image" (Avraham, 2015). The model analyzes three characteristics of marketing strategies: the type of crisis, the location, and the target audience. After initial analysis, this is followed up with three types of marketing strategies: focusing on the source of the negative message, the target audience, and the message itself. Source strategies influence or replace the source believed to spread negative information on the destination. Audience strategies are concerned with the audience's values, perceptions, and dreams. Finally, message strategies focus on contradicting the negative messages, stereotypes, perceptions, and generalizations spread about the destinations (Avraham & Ketter, 2008a).

CASE STUDY OF THE TOHOKU REGION

The Tohoku Region refers to the six northern Japanese prefectures: Aomori, Akita Yamagata, Iwate, Miyagi, and Fukushima, which are renowned for its cold climate, hot springs, and pristine scenery. However, the region has long been in decline in both its population and its economy. Prior to the Second World War, the regional economy was focused on agricultural and industrial activities such as livestock raising, food production, fisheries, and steel works. However, during the 1980s, the steel industry went into decline, partially due to the rising value of the yen. In addition, the region's role as a food supplier to Japan's metropolitan area also decreased as import barriers were reduced leading to a rise of imports of agriculture and livestock from Australia, China, South East Asia, and the United States (Fujita & Hill, 1993).

The 1970s and 1980s brought about numerous changes to the accessibility of cities in the Tohoku Region. The creation of an expressway, *shinkansen* rail lines, and construction of new airports have contributed to changes in the type of economic activities. Manufacturing firms moved manufacturing into the rural areas of Japan to take advantage of cheaper rent, lower labor costs, and improved transportation access (Okawa, 2005). With the decline of agriculture, many farmers found part-time or seasonal employment in these factories as they may work in the farms at certain periods, then switch to other industries for employment. Automotive companies such as Toyota and electronic companies such as Sony and Toshiba maintain plants in Iwate and Miyagi (McGee, 1991). Despite attempts to adapt to the changing global economy and improvements in transportation linkages, Tohoku's population continued to decline.

To combat the drop in population and job losses, the national government introduced the 1987 Law for Development of Comprehensive Resort Areas, also known as the Resort Law. Through this plan, the government sought to revitalize the decaying economies of Japan's peripheral regions. The law divided Japan into categories of potential tourism destinations utilizing the unique geographic traits of each region. By promoting incentives for resort development, the law sought to creatie employment opportunities, reduce outmigration of young labor, and as a result, increase tax revenues (Rimmer,

1992). Within the Tohoku Region, six resort regions were carved out: the Tsugaru-Iwaki Resort Plan in Aomori, the 40 Degree Seasonal Resort Akita Plan in Akita Prefecture, the Zao-Gassan Regional Resort Plan in Yamagata Prefecture, the Sanriku-Rias Resort Plan in Iwate Prefecture, the Kurikoma-Funagata Refresh Resort Oasis 21 Plan in Miyagi Prefecture, and the Aizu Fresh Resort Plan in Fukushima Prefecture (Fig. 1).

Despite the ambitious goals of the Resort Law, its ability to change the tide of decline was limited. Funck assesses the problems in the Resort Law's planning, which have led to its failure. These problems include the scale of the development, which was done at a national level but failed to consider local resources and input from local stakeholders. In addition, the law attempted to leave development to market forces but yet allowed every prefecture to submit similar numbers and types of bidding. As a result, neighboring prefectures were offering similar products with each other, and many construction plans were either economically unviable or never realized, especially after the effects of the bubble economy became apparent (Funck, 1999).

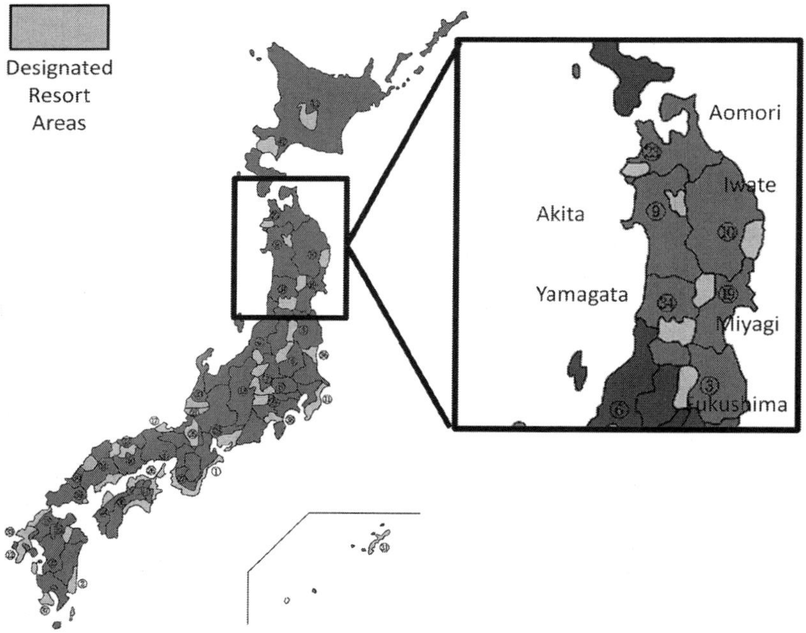

Fig. 1. The Resort Law's Designated Resort Zones. (Modified from the Japanese Ministry of Land, Infrastructure, Transport and Tourism.)

In 2011, the largest earthquake ever recorded in Japan, struck on March 11 at 14:46 local time. The combination of the earthquake and resulting tsunamis led to over 15,800 deaths, over 2,500 missing, over 6,100 injured, and more than one million buildings that ranged from being completely destroyed to partially damaged. In total, over 470,000 people were evacuated from their homes and as of 2016, 50,000 are still residing in temporary housing. In addition to the natural disasters, the earthquake and tsunami triggered a nuclear accident at a nuclear power plant in eastern Fukushima, leading to the evacuation of roughly 154,000 residents from the restricted areas. Direct financial damages are estimated to be approximately 16.9 trillion yen or 100 billion US dollars. Of the six Tohoku Prefectures, Aomori, Iwate, Miyagi, and Fukushima Prefectures suffered the most damages and casualties due to their geographic proximity to the epicenter of the earthquake (Japan Reconstruction Agency, 2016).

The events of 2011 sent an already ailing economy further downward. The number of tourism arrivals immediately dropped after the disasters. As of 2016, five years after the disaster, none of the four worst hit Tohoku Prefectures have seen annual in-bound tourism numbers recover to pre-disaster levels (Fig. 2). Aomori and Iwate Prefectures' tourism growth are stagnating, while Miyagi and Fukushima Prefectures' numbers continue to

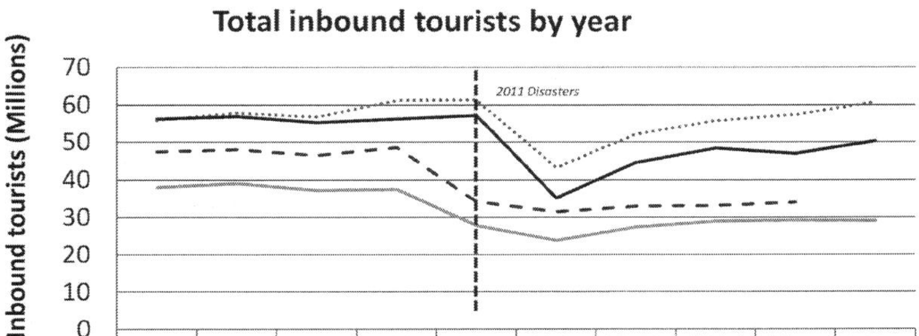

	2006	2007	2008	2009	2010	2011	2012	2013	2014	2015
- - Aomori	47.4	48	46.4	48.5	34.2	31.5	33	33.2	34	
—— Iwate	38	39	37.2	37.5	27.9	23.8	27.4	28.9	29.2	29
······ Miyagi	55.8	57.9	56.8	61.2	61.3	43.2	52.1	55.7	57.4	60.7
—— Fukushima	56.2	56.9	55.3	56.2	57.2	35.2	44.5	48.3	46.9	50.3

Fig. 2. Annual Changes in In-bound Tourists Since 2006. Note Aomori Prefecture Adjusted Its Method of Counting Tourists, Leading to the Significant Drop Between 2009 and 2010. (Compiled from Aomori, Iwate, Miyagi, and Fukushima Prefectural Statistical Yearbooks.)

climb. It is expected that Miyagi Prefecture will be the first to recover its annual tourism numbers to pre-disaster levels at by 2017–2018.

METHODOLOGY

Avraham and Ketter identify two types of media strategies that destinations use in restore its image—cosmetic and strategic (Table 1). The first are cosmetic approaches, which are mostly focused on changing the negative image without changing the reality that caused the image problem. Second are strategic approaches that attempt to change both the image and the reality through comprehensive actions and campaigns that lead to substantial changes on a destination's reality. Because of its nature, strategic approaches generally involve the inclusion of residents, vision formation, strategy design, defined long- and short-term goals, enhancing local pride, employing research tools, and target audience analysis (Avraham & Ketter, 2013). This strategic versus cosmetic approach serves as the basis for this study's methodology.

This study asks the following research questions in order to determine the strategies used by Tohoku's Prefectures to alter a more positive image of their prefecture in order to recover from tourism decline:

1. What are the problematic images affecting tourism in the prefecture?
2. What kind of marketing strategies are being used to combat these images?
3. To what extent do these marketing strategies address the root problems? Are the approaches being used seek a cosmetic change or an in-depth and more strategic change?

To answer these questions, a qualitative research methodology is used. Qualitative content analysis is considered as an effective method to analyze promotional campaign. Texts can show general discourse patterns through the appearance of labels, symbols, and slogans found in promotional material (Pauly, 1991). As such, the methodology for this research relies on the collection of documents and participant interviews. Background data such as tourism statistical data are collected from primary sources such as prefectural and municipal statistical yearbooks. Data on stereotypes and images of the Tohoku Region are collected from secondary data, such as books, academic articles, and other literature that have explored this phenomenon. In addition, interviews with tourism also sought to confirm these stereotypes by market segment.

Table 1. Avraham and Ketter's Theoretical Model on Media Strategies for Destinations Suffering from Prolonged Negative Images. (Modified from Avraham and Ketter, 2013.)

Levels of Change in Destination's Characteristics	Approach	Strategy	Description
Low	Cosmetic Approach	Ignoring	Marketers pretend nothing is wrong with their destination image despite suffering from stereotypes, and markets other aspects such as its past or future
		Disassociation	Identifies an associated negative image and tries to disassociate themselves from this problematic location to avoid the negative image
		Association	Strategy that seeks to link a destination to another destination with a favourable image
		Acknowledging	Acknowledges the negative image directly, but that it has been improved and now attractive again
		Counter-Message	Delivering counter-messages aiming to battle the negative perceptions
		Spin-Message	Spin a destination's negative aspects into an asset
		Ridiculing	Using humor to break stereotypes by making negative image absurd
	Strategic Approach	Hosting Events	Spotlight events that don't necessarily encounter roots of the image problem but can broaden its image. However it can change the reality only in the short term and its impact not long lasting
		Spinning Events	By recognizing the negative factors responsible for the image, spins it into a positive trait. In every bad thing, there is also good
		Tackling	Directly address the reasons that caused the problematic perception
High		Counter-Branding	A total and in-depth change in the destination characteristics leading to the greatest extent of change. It is branding counter to its stereotype.

Next, to examine marketing strategies to combat negative images, two methods were used. First, prefectural tourism development plans were collected directly from tourism and planning department websites. These plans provide information on how the prefectures guide tourism development for the entire prefecture and cities that are key destinations for tourists. Next, secondary data, such as official tourism pamphlets, tourism webpages, and other promotional media were collected for content analysis.

Second, interviews were conducted with tourism stakeholders from Aomori City, Morioka, Sendai, and Fukushima City, which represent the capital cities of the four case study prefectures and also serve as the main transportation hub that links visitors to nearly all other attractions within their respective prefectures. Interviews were either conducted in person or over email. The interviews are semi-structured and seek to identify the target audience of the media campaigns, characteristics of these audiences, initiatives aimed toward these audiences, and the successes and challenges thus far.

FINDINGS

Tohoku as a Stereotype

The Tohoku Region has been characterized as rural, mountainous, and suffering from economic decay and depopulation. Its inhabitants are stereotyped as uneducated farmers, poor, and backward. However, at the same time, they are also romanticized as being unsullied, hardworking, and willing to work for less than their counterparts in the more urban and populous regions of Kanto and Kansai (Brown, 2006; Kumagai, 2011; Traphagan, 2000). The triple disasters of 2011 further tarnished the regional image as both national and international mass media continue to portray images of destruction and concerns over lingering radiation issues, which has been associated to the entire region rather than limiting it to afflicted areas (Dinitto, 2014; Rausch, 2012; Tran, 2015).

Case Study: Aomori Prefecture

Aomori Prefecture is the northernmost prefecture in the Tohoku Region, as well as the northernmost prefecture on Honshu Island. Unlike the other prefectures of Tohoku, Aomori Prefecture borders both the Sea of Japan and the Pacific Ocean, and the northern coastline is surrounded by the Tsugaru Strait that separates them from Hokkaido. Of the four prefectures examined in this study, Aomori Prefecture suffered the least in terms of human casualties and physical damage, from the 2011 earthquake and tsunami. Direct damages were mostly limited to the prefecture's eastern coast, particularly in Hachinohe and Misawa. Despite limited physical damage, in-bound tourism declined after the disasters (Fig. 3).

To address the decline in tourism, the prefecture published the *Aomori-Ken Kanko Fukkou Taisaku Jisshi Keikaku* or Aomori Prefecture Tourism Recovery Counter Measure. The plan acknowledges the effects of harmful rumors to in-bound tourism, especially among foreign tourists. To combat these negative images, the prefecture seeks to promote more media promotion and exchanges with domestic and international markets (Aomori Prefecture, 2016).

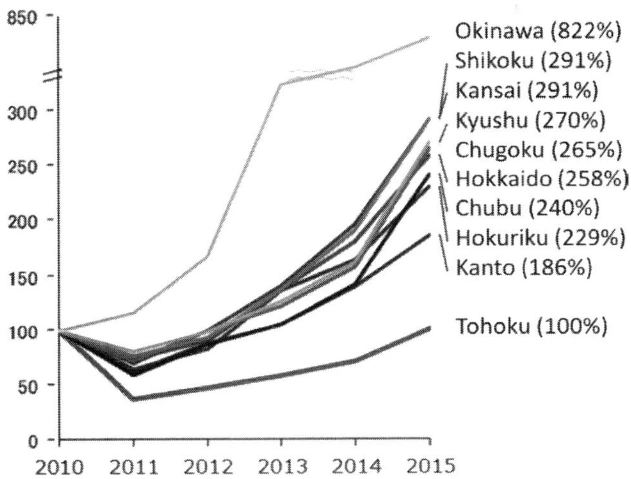

Fig. 3. Changes in the Ratio of Foreign Tourists Since 2010. (Modified from Aomori Prefecture 2016.)

The plan highlights Aomori's numerous physical attributes, such as its four seasons and natural scenery, which include Lake Towada and Shirakami Sanchi, one of Tohoku's two United Nations Educational, Scientific and Cultural Organization (UNESCO) World Heritage sites. Cultural attributes are also heavily promoted such as the Nebuta Festival held in August, local hot springs, and historical sites such as Hirosaki Castle. It should be noted that much of the attractions being promoted in Aomori are either inland or on its western coast.

Media strategies toward domestic tourists differ from foreign ones. For domestic tourists, Aomori Prefecture seeks to spin its stereotype as a backward, rural, and cold hinterland into Japan's most traditional and historic area. The messages play into the nostalgic feelings of domestic tourists from major metropolitan areas by portraying Aomori as the remaining refuge of Old Japan. In addition, the cold weather is also spun as an opportunity for a variety of winter activities.

The promotion of foreign tourists is seen as an important element in economic recovery. In all regions in Japan, the share of foreign tourists in each region has exponentially increased. The number of incoming foreign tourists is expected to further increase by the time of the 2020 Olympics. Despite the increase in foreign tourism numbers and relaxed visa policies, the overall growth in the Tohoku Region remains stagnant while other regions continue to experience significant foreign tourism growth (Fig. 3). Aomori's recovery plan acknowledges that more qualitative research needs to be done to better understand correlations among age, gender, and interests of overseas tourists, as awareness of Aomori is relatively low.

Among the foreign visitors, Aomori seeks to target Taiwanese tourists who are the most common foreign group, followed by South Koreans, Chinese from prosperous cities such as Shanghai and Guangzhou, Australians, Thais, and Malaysians. In the cases of foreigners who hail from warm climates such as Guangzhou or Australia, Aomori seeks to emphasize its snowy climate. For all foreign groups, the prefecture aims to promote itself as Japan's best kept secret allowing foreigners to experience a more authentic and traditional Japan that may have been lost in other more populated regions.

The prefecture's tourism promotional media for tourists, such as its official tourism websites, pamphlets, and guides, tend to avoid the use of the term Tohoku. Instead, its geographical proximity to the Tsugaru Strait and Hokkaido are emphasized, especially in relation to the 2016 commencement of high-speed rail access from Tokyo to Hokkaido, in which Aomori has become the main transfer point. The prefecture and municipalities have been in close coordination with their counterparts in Hokkaido to maximize tourism development in both areas.

By disassociating itself with the Tohoku Region and associating itself with the more prosperous and well-known Hokkaido area, Aomori Prefecture's strategy to alter negative images has mostly been cosmetic. The prefecture's physical tourism attributes have not changed, and instead, the prefecture seeks to spin the negative images of ruralness and decline into images of undiscovered, pristine nature, traditional lifestyles and architecture reminiscent of the Edo Period, home-to-world renowned sites, and abundant opportunities for winter sports. Concerns over stereotypes of radiation and disaster are ignored.

Case Study: Iwate Prefecture

Iwate Prefecture is Japan's largest prefecture on Honshu Island, and the second largest prefecture in the country, behind Hokkaido. Iwate is located south of Aomori Prefecture and north of Miyagi Prefecture. Despite its large size, the prefecture is mostly mountainous with much of the prefecture's population residing in the central valley, alongside the Kitakami River, or on its eastern coastline facing the Pacific Ocean. Along with Miyagi and Fukushima

Prefectures, Iwate was severely impacted by the 2011 Great East Japan Earthquake and Tsunami, with over 4,600 deaths and over 1,600 missing.

In 2014, the *Michinoku Iwate kankorikken dai ni ki kihon keikaku no sakutei* or the Second Michinoku Iwate Tourism Basic Plan was published. The plan is the second and most recent of a series of tourism recovery plans, which seeks to recover inbound tourism numbers to pre-disaster levels. Although tourism numbers have increased, as of 2016 it has still not reached pre-disaster levels and have entered a stagnation phase (Fig. 3). The prefecture's coastal cities have seen the biggest decline in tourism. Current tourism statistical data indicate that nearly half of in-bound tourists visit southern Iwate where UNESCO World Heritage site, Hiraizumi, is located. Another 35% visit the central part of Iwate where capital city Morioka is located, 12% visiting the coastal cities, and 10% to the north. In terms of activities, tourists are seeking consumption of local food, hot springs, and events and festivals. Although Iwate has abundant snowfall in winter, sports and recreational activities were ranked the lowest among tourists (Iwate Prefecture, 2014).

The plan acknowledges that anxiety and fears over radioactive contamination are factors in the decline of tourists into the prefecture. Municipal tourism representatives in Morioka and Miyako have stated tourists either perceive Iwate as risky or have completely no knowledge of the region at all. To counter these stereotypes, the tourism plan seeks to improve awareness of Iwate through educational tours of disaster prevention, provide accurate information, and develop a strong campaign in the Kanto Area, which includes Tokyo.

As with Aomori, the potential of foreign tourists is being strongly considered because the prefecture seeks to promote increased tourism numbers among South Koreans, Chinese, Taiwanese, Americans, Australians, and Thais. The recovery of Taiwanese tourists in particular is considered important for economic recovery. There are plans to improve the availability of multilingual information, increase transportation linkages, and increase exchanges with East Asian Countries.

As with other Tohoku Prefectures, Iwate Prefecture also faces ongoing problems with its enduring image as a cold, underdeveloped part of Japan, with few amenities and activities for visitors. Hiraizumi continues to be a central part of the plan due to its status as a World Heritage Site. In addition, the prefecture seeks the promotion and development of the Reconstruction National Park and the Sanriku Geopark in its coast. Hosting events such as athletic meets are key to improving the prefecture's appeal for sports activities. Perhaps most importantly, the plan seeks the continued broadcast and distribution of the NHK Drama, Amachan, which is set in the Sanriku coastal areas of southern Iwate. The drama has served as an important tool in strengthening awareness of Iwate to domestic audiences. In terms of foreign tourists, the prefecture is strengthening its use of Internet and social media tools to spread awareness of Iwate's attributes and to improve brand recognition.

Promotional tourism media from Iwate Prefecture were also analyzed, including official tourism websites, pamphlets, and other guides. Similar to Aomori, disassociation strategies were found as the prefecture limits or completely avoids the use of the term Tohoku. Association strategies were also found as guidebooks liken Mount Iwate as the northern version of Mount Fuji. Interviews with a municipal tourism employee discovered plans to address radiation concerns, such as explaining Iwate's location as being much farther from Fukushima than Tokyo; however, current tourism media ignores addressing radiation concerns. Anxieties over the images of destruction are addressed with information being provided on the recovery progress, as well as expressions of gratitude to those who contributed to the prefecture's recovery. It is worth noting that the statistical data state that 8% of tourists come to Iwate to examine disaster sites.

Currently Iwate's approach toward countering negative images rely heavily on cosmetic approaches by disassociating itself from Tohoku and instead simply highlighting its two key areas, Hiraizumi and the Sanriku Coast. Destruction stereotypes are also acknowledged, as tourism media accepts that it is an issue but that progress has been made and that the area is becoming an attractive destination. The prefecture's tourism plans, however, indicate a more strategic approach as it seeks to host a variety of conferences and sport events to counter its underutilized potential as a winter sports location. More significantly, Iwate plans to develop and implement a counter-branding approach to the Sanriku Region, changing its perception as an area of devastation to one of recovery, resilience, and preservation.

Case Study: Miyagi Prefecture

Miyagi Prefecture is the most populous prefecture of the Tohoku Region and contains Sendai, the most populous city in Tohoku. Located south of Iwate Prefecture, and north of Fukushima Prefecture, Miyagi is unique in its mostly flat topography, although its western regions bordering Yamagata Prefecture are mountainous. The prefecture suffered the highest casualties caused by the 2011 Great East Japan Earthquake and Tsunami, including over 9,400 deaths, and over 2,000 missing.

The 2014 *Dai San Ki Miyagi Kanko Senryaku Puran* or Third Miyagi Prefecture Tourism Strategic Plan seeks to restore tourism to pre-disaster levels. Currently, out of the four prefectures examined, Miyagi Prefecture is the closest in reaching pre-disaster tourism numbers. It is expected to surpass 2010 numbers by 2017 or 2018. Tourist concerns over destruction and radiation also persist in Miyagi and are believed to be one of the main deterrents for tourists entering the region. Gaps also persist between inland and coastal tourism as only 8 million of its 57 million tourists visit the latter areas (Miyagi Prefecture, 2014).

Tourism infrastructural plans for Miyagi include the expansion of capabilities of Sendai Airport, increasing flights to China, South Korea, Taiwan, Hong Kong, Thailand, Singapore, and Malaysia. Improvements in these areas not only affect Miyagi Prefecture but also the Tohoku Region as a whole as Sendai Airport is the closest international airport for neighboring Fukushima, Iwate, and Yamagata Prefectures. Physical attributes include Matsushima, traditionally regarded as one of Japan's three great views – hot spring towns of Naruko, Akiu, and Sakunami – and the mountainous areas of Zao, Shiroishi, and Kurikoma. The plan also seeks to improve linkages between inland and coastal tourism, specifically Ishinomaki and Kesennuma cities.

Roughly half of the tourists in Miyagi are internal, about a quarter from the Kanto region, 16% from other Tohoku Region, and 1% from the Chubu and Kansai regions of Japan. To attract more tourists from other prefectures, exchange of awareness campaigns are being pushed, especially in the Kanto area, which includes Tokyo. For the Chubu and Kansai regions, a new campaign "I want to go to Tohoku" is being promoted to create a new market segment.

Increasing foreign tourists is also acknowledged to be a key factor in recovery. Key tourism markets are Taiwan, China, South Korea, and Hong Kong. In addition, Southeast Asian markets such as Thailand, Singapore, and Malaysia are expected to increase. Although comparatively fewer, American and European tourists are also considered important markets. However, Taiwanese and Hong Kong tourists are considered the most important market as they are more likely to be repeat visitors.

In an interview with Sendai's Municipal Tourism Department, South Korean tourists are currently the hardest to recover due to persisting images of destruction and fears over radiation. On the other hand, it is believed that Taiwanese tourists are better informed of local conditions as well as local attributes, leading to increased repeat visits. While Tokyo and Kyoto are the two most common destinations for foreign tourists, repeat visitors will often visit Sendai on the next visit. Western tourists, including Europeans and

Americans, are the most challenging market as they have very low awareness of Miyagi and Tohoku as a whole. Thus strategies focus on popular media sites, such as Facebook and Twitter, and invite prominent bloggers to help create better awareness of Miyagi and Sendai's attractive aspects, leading to improved brand recognition.

To change the characteristics of the prefecture's negative images, Miyagi is unique in which it directly acknowledges all of its negative aspects. Fears over radiation are directly addressed on both the prefectural and Sendai City's tourism webpages by providing radiation readings, and delivering a counter-message by comparing these readings with other major cities such as Paris demonstrating that the prefecture's radiations are not only normal but also lower than that in other areas of the world. Fears over destruction are also acknowledged through the promotion of disaster safety events in its coastal areas. More strategically, the prefecture has been aggressive in hosting sporting events such as baseball, marathons, and the upcoming 2020 Tokyo Olympics where the governor is pushing for Miyagi to become the venue for the canoe and rowing events.

Miyagi Prefecture is unique in that the cosmetic approaches it uses are significantly different than those used in Aomori and Iwate. Rather than ignoring problems and disassociating itself from Tohoku, the prefecture acknowledges concerns over radiation and destruction and offers counter-messages to assure visitors that Miyagi is much safer than stereotypes portray it to be. More strategically, the prefecture has developed a number of spotlight events that seek to broaden Miyagi's image as a sport and cultural center. The prefecture also attempts to tackle the perception of its coastal cities as being dangerous and devastated by demonstrating improved safety features for tourists. The prefecture's tourism plan hopes to reach 67 million tourists by 2017. Although it is unknown whether the prefecture will indeed reach this target, it is the best positioned in the Tohoku Region to reach pre-disaster levels.

Case Study: Fukushima Prefecture

Fukushima Prefecture is the southernmost prefecture in the Tohoku Region, located south of Miyagi Prefecture and north of Tochigi, Gunma, and Ibaraki Prefectures in the Kanto Region. Fukushima Prefecture is also the second most populous prefecture in Tohoku, with its largest city being Koriyama rather than the prefectural capital of Fukushima City. The prefecture was devastated by the 2011 Great East Japan Earthquake and Tsunami, with over

1,800 deaths and 119 missing. Following the tsunami, active reactors in Tokyo Electric Power Company's (TEPCO) Fukushima Daiichi Nuclear Power Plant were shut down. However, the destruction of the emergency generators prevented the plant's ability to cool the reactors, leading to three nuclear meltdowns and the release of radioactive material. Concerns over radiation led to the creation of evacuation zones in the coastal areas of Fukushima near the power plant. Towns such as Futaba and Namie are now part of an exclusion zone, with the majority of its population gone but neighboring areas such Minamisoma City and Naraha are experiencing significant depopulation.

Currently, there exists no dedicated tourism recovery plan by Fukushima Prefecture. However, a section dedicated to tourism recovery exists within a larger economic recovery plan. In addition, a number of cities such as Fukushima City and Aizu-Wakamatsu have developed their own tourism recovery plans. For the sake of this study, the 2014 *Fukushima Kanko Keikaku Chosa Bunseki* or Fukushima Tourism Plan Survey Analysis was used as a reference for the prefecture's tourism development goals.

Radiation concerns dominate the plans as it is the primary contributor to the large population decrease, and the major detriment of all products from Fukushima as consumers are afraid to purchase any local goods or visit the area. Both domestic and foreign tourists have declined. Although the prefecture has regained some of its losses since 2011, tourism numbers appear to be stagnating (Fig. 4). To dispel the negative images associated to the prefecture, the prefecture seeks to cooperate with the Japan National Tourism Organization (Fukushima Prefecture, 2014).

To increase tourism, the plan seeks to develop a mascot character, cooperate with idol groups, improve its portrayal in media and film, and work with travel agencies to improve Fukushima's image. Like Iwate, Fukushima is hoping the continued broadcast of the NHK Drama, "Yae no Sakura" starring popular actress Haruka Ayase, could spread awareness and improve the perception of the prefecture among domestic audiences. The prefecture believes it has valuable tourism assets such as its rich nature, castles, industrial heritage, history, and cuisines, but which are not well known.

Tourism media from Fukushima Prefecture have opted for a primarily cosmetic approach, seeking to ignore its current situation, and instead focus on its future, with slogans such as "Future From Fukushima," "Towards a New Future," "Forming a Link to the future, Beautiful Fukushima." In addition, much of its past are also emphasized, particularly its location as a major battlefield during the Sengoku Period. Focusing on images of the past and future are common among areas currently experiencing prolonged negative images

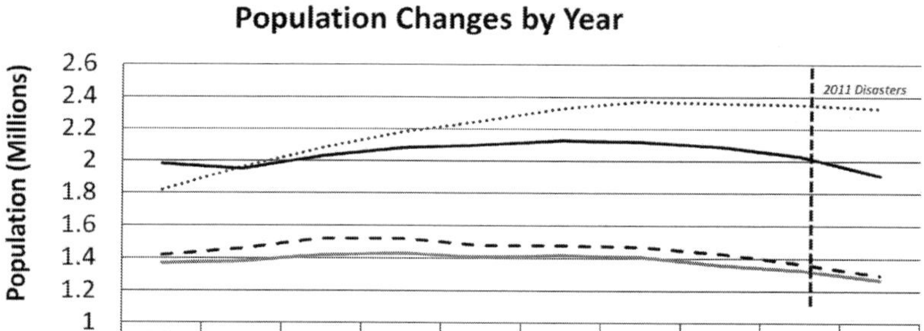

	1970	1975	1980	1985	1990	1995	2000	2005	2010	2015
- - Aomori	1.42	1.46	1.52	1.52	1.48	1.48	1.47	1.43	1.37	1.3
—— Iwate	1.37	1.38	1.42	1.43	1.41	1.42	1.41	1.36	1.33	1.27
...... Miyagi	1.82	1.96	2.08	2.18	2.25	2.33	2.37	2.36	2.35	2.33
—— Fukushima	1.98	1.95	2.03	2.08	2.1	2.13	2.12	2.09	2.03	1.91

Fig. 4. Annual Changes in Population Since 1970. (Compiled from Aomori, Iwate, Miyagi, and Fukushima Prefectural Statistical Yearbooks.)

and stereotypes and have yet to address them (Table 2). There has been no mention of marketing strategies for different tourism markets or other cosmetic and strategic approaches to combat Fukushima's negative images beyond hosting spotlight events. Similar to Miyagi Prefecture, Fukushima's official tourism portal also provides radiation levels and attempts to downplay its overall effects by comparing it with select world cities.

CONCLUSION: WHAT'S IN A NAME?

This chapter presents four different case studies on how Tohoku's most devastated regions have attempted to recover its tourism numbers and combat the negative images associated with the 2011 disasters as well as pre-existing stereotypes of a region that is cold, backward, and empty. The different approaches used by each prefecture also reflect the differences in the severity of the disasters. As Aomori Prefecture is the furthest from the earthquake's epicenter, it received comparatively less casualties and damages. In addition, Aomori is located on the periphery of Tohoku, far enough from Fukushima, which allows them to disassociate from the region, and instead, focus on associating itself with Hokkaido, which is currently undergoing rapid tourism growth.

For the Prefectures of Miyagi, Iwate, and Fukushima, ignoring the damages caused by 2011 disasters may perhaps be unavoidable as both Miyagi

Table 2. Destination Branding Characteristics.

	Aomori	Iwate	Miyagi	Fukushima
Brand Slogans	• Feel Aomori, Feel Japan • Japan's Best Kept Secret • A new discovery in every visit	• Experience Japan's Heartland • Have a Nice Trip	• Journey to Miyagi • Sendai and Miyagi, where Smiles Blossom • I want to go to Tohoku!	• Future from Fukushima • Future from now • Towards a new future • Forming a link to the future, beautiful Fukushima
Physical attributes	• UNESCO: Shirakami • Nature • Four Seasons • Hot Springs • Food: Apples • Castles • Festivals: Nebuta • Winter Sports	• UNESCO: Hiraizumi • Nature • Four Seasons • Hot Springs • Food: Noodles • Winter Sports	• Nature • Japan's 3 Scenic Views: Matsushima • Castles • Hot Springs • Food: Cow Tongue	• Nature • Food: Cherries • Castles • Culture • Winter Sports
Emotional attributes	• Unspoiled • Nostalgic • Outdoor Adventure • Undiscovered	• Charming • Nostalgic • Outdoor Adventure	• Unspoiled • Modern • Sophisticated • Nostalgic • Outdoor Adventure	• Charming • Nostalgic • Educational
Image approach	• Disassociation • Association • Spin Message	• Disassociation • Association • Hosting Events	• Acknowledging • Tackling • Hosting Events	• Acknowledging • Disassociation • Ignoring

and Iwate acknowledge its impact. Of the four case studies, rather than ignoring or disassociating itself from problems, Miyagi Prefecture has been the most forthright in acknowledging and addressing its negative stereotypes. It transforms concerns over radiation and destruction into normalcy and safety by explaining to tourists that radiation levels are lower than many of the world's major cities and informing tourists of improved disaster safety in its coastal cities. In addition, Miyagi Prefecture is set to be the first Tohoku prefecture to return to pre-disaster inbound tourism levels.

Fukushima Prefecture faces the most arduous task of overall recovery due to ongoing concerns over radiation. Unlike Aomori, which can link itself to other areas, and Miyagi persuading people that radiation levels are safe,

Fukushima Prefecture contains a number of no-entry zones that have turned populated areas into ghost towns. The disassociation tactics used by Aomori and Iwate also highlight another problem Fukushima faces, the name of the disaster. Fukushima Nuclear Disaster, Fukushima Accident, and Fukushima Nuclear Meltdown are just one of the many names used to refer to the 2011 meltdown. Rather than referring it as the TEPCO accident, which would acknowledge TEPCO's role in the meltdown, Fukushima has emerged as the standard name for the accident. The association of Fukushima, rather than TEPCO, to the meltdown has damaged the reputation of the entire prefectures, including areas such as Aizu-Wakamatsu, which is located far from the problematic areas.

A review of the tourism recovery plans of the four case studies also identified important areas for further research. As Aomori, Iwate, Miyagi, and Fukushima seek to improve name brand and spread awareness of its natural, cultural, and built assets to domestic and foreign tourists, the actual tourism product and messages are roughly similar. Physical assets, such as pristine nature, hot springs, four seasons, snow and winter activities, and outdoor recreation, while seeking to play onto nostalgic sentiments among domestic tourists, and an unspoiled, authentic Japanese experience to foreigners, were common in all four prefectures (Table 2). With little in terms of product differentiation in conjunction with the four prefectures focusing on the same tourism markets, particularly the Taiwanese, tourists will find it hard to justify visiting multiple Tohoku locations if one area offers the same things. Aomori Prefecture's tourism acknowledges the need for a Destination Marketing Organization (DMO), which would better coordinate and nurture local tourism. Interviews with various Miyagi Prefecture as well as municipal tourism departments in Miyagi and Iwate also echoed the need for a DMO, which would improve private – public collaboration and contribute to a more cohesive regional tourism development plan.

Future research into destination image repair that utilizes all elements of Avraham and Ketter's multistep model can be utilized to study in further detail the approaches each prefecture and municipality is using to alter place image. While this study focused primarily on the message, the multistep model also includes further research on the characteristics of the audiences receiving the message, such as their values, perceptions, and dreams. In addition, the role of regional DMOs can also provide insight on how places can create a comprehensive and integrated tourism plan where competition between destinations in the same region are limited. Within Japan, the only regional DMO that exists is found in Okinawa Prefecture. An analysis of how Okinawa's DMO manages product differentiation for each of Okinawa's island could provide useful information for Tohoku to develop a regional framework that would create incentives for tourists to visit multiple areas.

ACKNOWLEDGMENTS

We would like to thank Sendai City, Matsushima Town, Miyagi Prefecture, Fukushima City, Aomori Prefecture, Miyako City, Iwate Prefecture, and Miyagi Minami DMO for their time to meet with us to explain the current situation in tourism recovery in the Tohoku Region.

REFERENCES

Anholt, S. (2006). Public diplomacy and place branding: Where's the link? *Place Branding*, *2*(4), 271–275.

Aomori Prefecture. (2016). *Aomori ken kanko fukko taisaku jisshi keikaku*. Aomori City: Tourism International Strategy Section Promotions and Exchange Division.

Avraham, E. (2015). Destination image repair during crisis: Attracting tourism during the Arab Spring uprisings. *Tourism Management*, *47*, 224–232.

Avraham, E., & Ketter, E. (2008a). Chapter 10 - The multi-step model for altering place image. *Media strategies for marketing places in crisis* (pp. 187–203). Oxford: Butterworth-Heinemann.

Avraham, E., & Ketter, E. (2008b). From theory to practice. *Media strategies for marketing places in crisis* (p. 99). Oxford: Butterworth-Heinemann.

Avraham, E., & Ketter, E. (2013). Marketing destinations with prolonged negative images: Towards a theoretical model. *Tourism Geographies*, *15*(1), 145–164.

Brown, L. K. (2006). Tohoku: A place. In C. S. Thompson & J. W. Traphagan (Eds.), *Wearing cultural styles in Japan: Concepts of tradition and modernity in practice*. New York: State University of New York Press.

Byrd, E. T. (2007). Stakeholders in sustainable tourism development and their roles: Applying stakeholder theory to sustainable tourism development. *Tourism Review*, *62*(2), 6–13.

Chanchai, T., Chao, Y., & Omthip, D. (2006). Effects of the December 2004 Tsunami and disaster management in Southern Thailand. *Science of Tsunami Hazards*, *24*(3).

Cioccio, L., & Michael, E. J. (2007). Hazard or disaster: Tourism management for the inevitable in Northeast Victoria. *Tourism Management*, *28*(1), 1–11.

Dinitto, R. (2014). Narrating the cultural trauma of 3/11: The debris of post-Fukushima literature and film. *Japan Forum*, *26*, 340–360. doi:10.1080/09555803.2014.915867

Faulkner, B. (2001). Towards a framework for tourism disaster management. *Tourism Management*, *22*(2), 135–147.

Fujita, K., & Hill, R. C. (Eds.). (1993). *Japanese cities*. Philadelphia: Temple University Press.

Fukushima Prefecture. (2014). *Kanko keikaku chosa bunseki*. Fukushima City.

Funck, C. (1999). When the bubble burst: Planning and reality in Japan's resort industry. *Current Issues in Tourism*, *2*(4), 333–353.

Huan, T.-C., Beaman, J., & Shelby, L. (2004). No-escape natural disaster: Mitigating impacts on tourism. *Annals of Tourism Research*, *31*(2), 255–273.

Iwate Prefecture. (2014). *Michinoku Iwate kankorikken dai ni ki kihon keikaku no sakutei*. Morioka: Commerce and Labor Tourism Department Tourism Division.

Japan Reconstruction Agency. (2016). *Efforts for reconstruction of Tohoku*. Tokyo: Government of Japan.

Kumagai, S. (2011). Tohoku dialects as a speech of rednecks: Language crossing in Japanese TV programs. *Studies in Humanities*, *61*(1–2), 153–169.

Lehto, X., Douglas, A. C., & Park, J. (2008). Mediating the effects of natural disasters on travel intention. *Journal of Travel & Tourism Marketing*, *23*(2–4), 29–43.

Lew, A. A. (2013). Scale, change and resilience in community tourism planning. *Tourism Geographies*, *16*(1), 14–22.

Mair, J., Ritchie, B. W., & Walters, G. (2016). Towards a research agenda for post-disaster and post-crisis recovery strategies for tourist destinations: A narrative review. *Current Issues in Tourism*, *19*(1), 1–26.

McGee, T. G. (1991). The emergence of Desakota Regions in Asia: Expanding a hypothesis. In N. S. Ginsburg, B. Koppel, & T. G. McGee (Eds.), *The extended metropolis: Settlement transition is Asia*. Honolulu: University of Hawaii Press.

Miyagi Prefecture. (2014). *Dai San Ki Miyagi Kanko Senryaku Puran*. Sendai: Economic Chamber of Commerce Tourism Department Tourism Division.

Nguyen, D., Imamura, F., & Iuchi, K. (2016). Disaster management in coastal tourism destinations: The case for transactive planning and social learning. *International Review for Spatial Planning and Sustainable Development*, *4*(2), 3–17.

Okawa, T. (2005). A note on the global challenges and local regional responses in the Tohoku District before the collapse of the bubble economy. *Bulletins of Yamagata University*, *35*(2).

Pauly, J. (1991). A beginner's guide to doing qualitative research in mass communication. *Journalism Monographs*, *125*(1–29).

Prideaux, B., Coghlan, A., & Falco-Mammone, F. (2008). Post crisis recovery. *Journal of Travel & Tourism Marketing*, *23*(2–4), 163–174.

Rausch, A. (2012). Framing a catastrophe: Portrayal of the 3.11 disaster by a local Japanese newspaper. *Electronic Journal of Contemporary Japanese Studies*, *12*(1).

Rimmer, P. J. (1992). Japan's resort archipelago: Creating regions of fun, pleasure, relaxation, and recreation. *Environment and Planning A*, *24*(11), 1599–1625.

Rosa, D. (2012). The Boxing Day tsunami and its effects on Thailand's tourism. *Neumann Business Review*.

Scott, N., Laws, E., & Prideaux, B. (2008). Tourism crises and marketing recovery strategies. *Journal of Travel & Tourism Marketing*, *23*(2–4), 1–13.

Shields, R. (1992). *Places on the margin: Alternative geographies of modernity*. London: Routledge.

Tavitiyaman, P., & Qu, H. (2013). Destination image and behavior intention of travelers to Thailand: The moderating effect of perceived risk. *Journal of Travel & Tourism Marketing*, *30*(3), 169–185.

Telfer, D. J. (2002). Tourism and regional development issues. In R. Sharpley & D. J. Telfer (Eds.), *Tourism and development: Concepts and issues*. Clevedon: Channel View Publications.

Tran, J. L. (2015). The problems and pleasure of publishing the horrors of the 3/11 tsunami. *The Japan Times*, May 7, 2015.

Traphagan, J. W. (2000). *Taming oblivion: Aging bodies and the fear of senility in Japan*. New York: State University of New York Press.
Wall, G. (2008). The Asian regional market: Flows and collaboration. In J. Cochrane (Ed.), *Asian tourism: Growth and change*. Amsterdam: Elsevier.
Walters, G., & Mair, J. (2012). The effectiveness of post-disaster recovery marketing messages: The case of the 2009 Australian Bushfires. *Journal of Travel & Tourism Marketing, 29*, 87–103. doi:10.1080/10548408.2012.638565
WTO (2010). *Manual on tourism and poverty alleviation: Practical steps for destinations*. Madrid: United Nations World Tourism Organization.

CHAPTER 4

RESTORATION OF COMMUNITIES FOLLOWING THE GREAT EAST JAPAN DISASTER: THE TRANSFORMATION OF MUTUAL HELP NETWORKS THROUGH THE EYES OF THE VICTIMS

Morio Onda

ABSTRACT

The earthquake and tsunami that struck eastern Japan on March 11, 2011, not only caused extensive direct damage to the population but also triggered a nuclear power plant accident that brought the terror and reality of radiation. The restoration of communities in Iwate, Miyagi, and Fukushima Prefectures presents enormous problems. People from the radiation-contaminated areas have faced numerous ordeals since resettlement after the accident. Through personal interviews with victims, this chapter investigates what happened in the regional societies and how community consciousness changed as a result of the combined natural and manmade catastrophes. The study focuses on the restoration

Recovering from Catastrophic Disaster in Asia
Community, Environment and Disaster Risk Management, Volume 18, 61–84

ISSN: 2040-7262/doi:10.1108/S2040-726220160000018010

of community from social bonds through mutual help networks as a spontaneous social order. As the result of interviewing, some propositions were developed concerning the transformation of mutual help networks. The stronger the outside assistance from volunteers whom the victims came to trust and rely on, the weaker inside communal help becomes. Inventorying and clarifying the particular problems of conflict in stricken communities such as the loss of confidence in neighbors, the possibilities of rebuilding communities are explored, especially indicating how to cope with the "social demise of communities" that local people had formed and occupied all their lives.

Keywords: Fukushima nuclear disaster; community consciousness; mutual help networks; spontaneous social order

INTRODUCTION

A great disaster struck eastern Japan on March 11, 2011. An extraordinary powerful earthquake (magnitude 9.0) and historically devastating tsunami (more than 10 m) destroyed houses and infrastructure and caused an estimated 15,889 deaths (National Police Agency) to Iwate, Miyagi, and Fukushima Prefectures (Fig. 1).[1] Then nuclear radiation, the "invisible terror," spread in Fukushima Prefecture and beyond, as the earthquake and the tsunami directly hit the Fukushima No. 1 (Daiichi) nuclear power station located on the seacoast in the center of its path (Fig. 2). While communities affected only by the earthquake and tsunami have been rebuilt physically and socially, it continues to be very difficult for people from the long-term radiation-contaminated areas to reconstruct their communities, as many have been unable to return to their former habitat at all. About 100,000 residents fleeing the radioactivity have dispersed far and wide throughout Japan. While many have started new lives in other areas of Japan, there are also many who remain in temporary housing unable to resettle in their former homes.

The social bonds among both groups of victims (those affected only by the natural disasters and additionally by the radioactivity) have changed. The general proposition that a disaster brings neighbors closer to each other does not necessarily hold in the face of truly devastating loss. Although some people have redeveloped ties in the quake-tsunami disaster areas, others, especially those whose houses were not destroyed by the tsunami, have

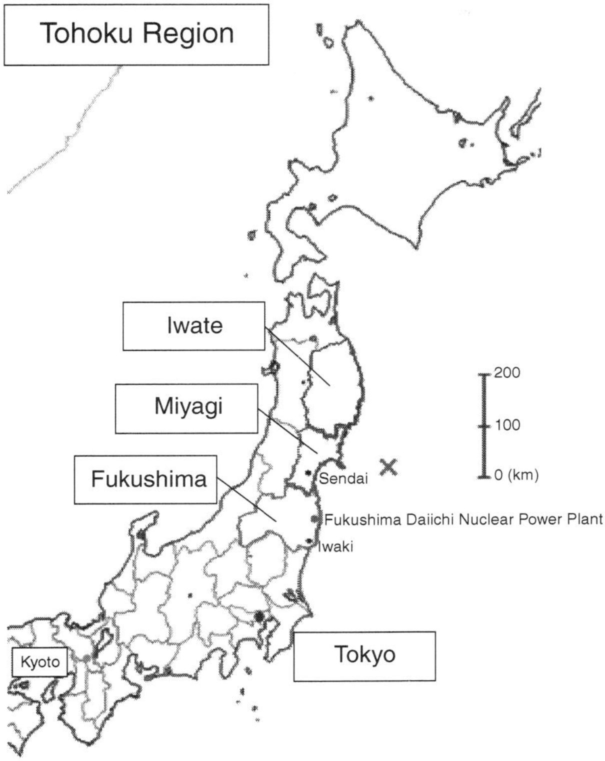

Date: March 11, 2011 at 14:46 JST
Epicenter: 130 km off the Pacific Coast of Tohoku District
Scale: Magnitude 9.0, Maximum Seismic Intensity 7
(The largest earthquake ever recorded in Japan)

Fig. 1. Scale of the Earthquake.
Source: Reconstruction Agency, Government of Japan (http://www.reconstruction.go.jp/english/topics/GEJE/index.html).

hesitated to communicate with less fortunate neighbors because of "survivors guilt." The stronger the outside assistance from volunteers whom the victims came to trust and rely on becomes, the weaker the inside communal help. This is a new proposition as a result of the research. The additional nuclear disaster inflicted quite another kind of damage on people. It was not only physical damage but also psychological damage. The results of

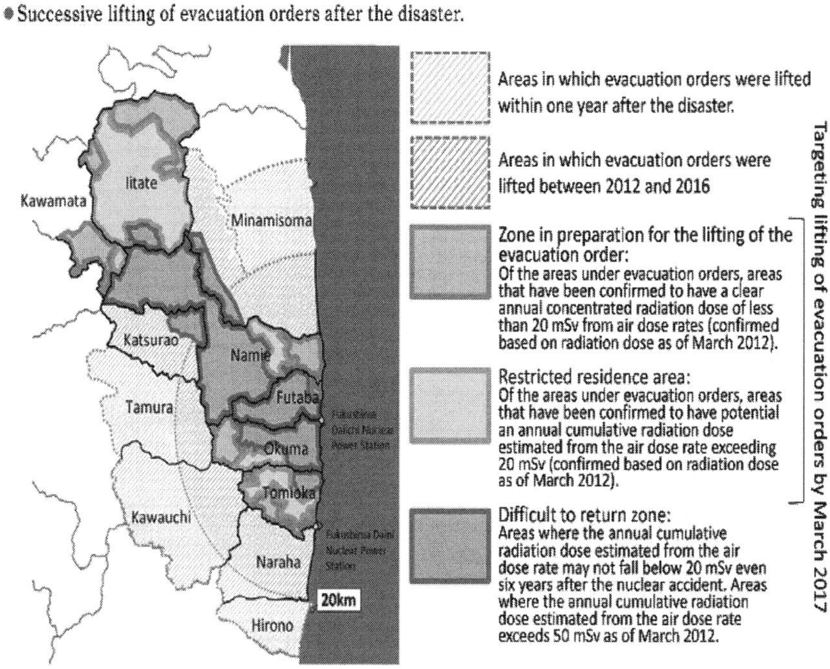

Fig. 2. Reconstruction and Recovery of Fukushima: Status of the Areas Under Evacuation Orders.
Source: Reconstruction Agency, Government of Japan (http://www.reconstruction.go.jp/english/topics/RR/index.html).

this study, which investigates this phenomenon for the first time, show that the stronger the previous ties of the victims were, the more fragile they felt when the bonds were materially fractured. These findings, not previously recognized, contribute to community development theory and practice. This study poses the following research questions: To what extent did the social bonds of traditional mutual help (reciprocity, redistribution, and unidirection) exist in the area at the time of the disaster? When victims were displaced, to what extent did each group maintain and lose the original bonds? If they were maintained and disrupted, why and how? How does each group view the desirability of restoring communities through these traditional bonds?

LITERATURE REVIEW

There are many surveys of natural disasters. Although some studies analyzed social and cultural impacts on the refugees from natural disaster, many of them focused only on the influence of the Indonesia Earthquake and Indian Ocean Tsunami (Boen & Jigyasu, 2005), Hurricane Katrina (Gardoni & Murphy, 2008), and Kobe Earthquake (Tatsuki, 2007; Chang, 2010). Local people were scattered by the disaster, so that solidarity was lost. People from the radiation-contaminated areas have faced numerous ordeals from the triple catastrophe. There has been little research into the social influence on community through nuclear disaster.

Social capital is said to be important in the recovering community (Hawkins & Maurer, 2009). What kind of social capital is indispensable after disaster? Superficial design of communities as outer layer by the central and local governments is not sufficient. The community with social capital and with a tradition of community activities could proactively participate in the reconstruction program, as reported in a study on the Kobe Earthquake in 1995 (Nakagawa & Shaw, 2004). In the reconstruction and rehabilitation phase, "Machizukuri Organization" that had community activities focusing on community betterment, health care of the elderly, voluntary services, and livelihood issues undertook the important activities. However, some local towns had other traditional activities in wedding or funeral ceremonies through mutual help networks in deep layer of communities. It is said that social relationship become strong after the disaster. Although there are some studies on community empowerment (Aldrich, 2012; Berke & Campanella, 2006), the support between the outside and the inside should be distinguished. Outside help sometimes brings local bonds into cracks in devastated community networks.

There have been some studies published since the Fukushima disaster. One of them is the research into creating an area management system based on "Machizukuri" (collaborative processes), where urban planning and other social systems (public administration and implementation of social policies) are more integrated (Murakami et al., 2014). The study of the community ownership of decision making in Miyako City of Iwate Prefecture and Kesennuma City of Miyagi Prefecture was done (Mochizuki, 2014). The research shows that "knit communities" whose members have acted of their own volition to help each other and educate themselves as to the reconstruction process from tsunami damage are important, and they are thus the intangible "asset" that makes these communities resilient. However, few sociological studies have been done on "refugee community" through mutual help networks.

METHODS

The study objective was to elucidate the transformation of mutual help networks in stricken communities to suggest new strategies for restoring communities after the enormous combined disasters (Fig. 3). The data were gathered through semistructured interviews that divided the victims into two broad categories: those who suffered only the natural disaster and those who suffered also from forced evacuation due to the danger of radiation (Wassermann & Faust, 2009). Interviewees were the victims who were willing to be interviewed in the refugee shelters. Because victims' attitude often changes according to the length of the shelter life, the timing of surveys is an important factor in conducting disaster research (Bourque, Shoaf, & Nguyen, 1997). Disaster recovery can be divided into four overlapping stages: (1) the emergency period; (2) the restoration period; (3) the replacement reconstruction period; and (4) the commemorative, betterment, and development reconstruction period (Haas, Kates, & Bowden, 1977). This study covers the first two stages, and interviews were conducted in the 12-month period following the disaster.

This chapter discusses the results of 36 post-disaster interviews of victims (Table 1). The author conducted interviews of the 12 victims affected only by the natural disaster in two cities each of Ibaraki, Iwate, and Miyagi

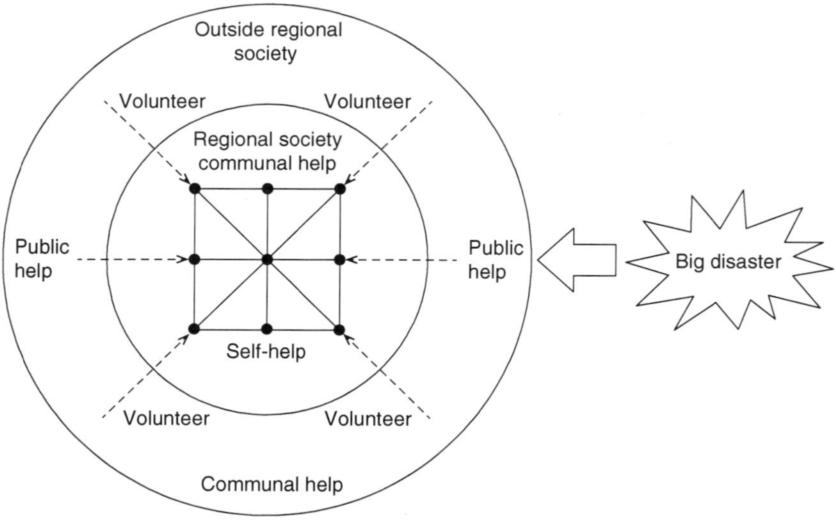

Fig. 3. Impact on Mutual Help Networks in Regional Society.

Table 1. Face Sheet of the Interviewed.

Type of Disaster	Residents of the Victims (City, Prefecture)	Generations Gender	Important Notice
Earthquake	Kitaibaraki-shi, Ibaraki	50s male	Voluntary refugee due to chronic disease
		70s male	Voluntary refugee due to wife's anxiety
		80s female	Voluntary refugee due to chronic disease
	Kamisu-shi, Ibaraki	40s female	Voluntary refugee as an apartment inhabitant
		50s female	Same as above
		60s male	Same as above
	Rikuzentakata-shi, Iwate	50s female	Avoided the tsunami due to living on high ground.
	Ofunato-shi, Iwate	40s male	Avoided the tsunami due to living on high ground.
	Kesennuma-shi, Miyagi	50s female	Avoided the tsunami due to living on high ground.
Earthquake Tsunami	Kitaibaraki-shi, Ibaraki	60s male	House was completely flooded by the tsunami.
	Ishinomaki-shi, Miyagi	70s male	House (first floor) was partially flooded by the tsunami.
	Kesennuma-shi, Miyagi	70s male	Scallop culture business
Earthquake Nuclear accident Fukushima Prefecture	Iwaki-shi,	70s female	House outside the designated area, but evacuated independently.
	Minamisouma-shi,	40s male	Tokyo Electric-affiliated company duty
		50s male	Full-time farmer
		70s male	Full-time farmer
	Okuma town	30s female	Tokyo Electric-affiliated company duty
	Futaba town	50s male	Tokyo Electric-affiliated company duty
		80s male	Former Futaba stationmaster
		30s female	Former Kisai High School refugee
		40s male	Same as above
		40s female	Same as above Town office temporary worker
		50s male	Former Kisai High School refugee Tokyo Electric-affiliated company duty
		60s male	Same as above
		60s male	Same as above
		60s male	Former Kisai High School refugee
		60s male	Same as above
		70s female	Same as above
		80s female	Same as above Former director of old men and women society
	Tomioka town	20s male	Care-related work

Table 1. (*Continued*)

Type of Disaster	Residents of the Victims (City, Prefecture)	Generations Gender	Important Notice
	Namie town	50s male	He was from Kesennuma and worked as a volunteer at the stricken area.
	Tamura-shi	50s male	After the emergency refuge preparations area was canceled, he came back.
		60s male	Same as above The head of a ward
Earthquake Tsunami Nuclear accident	Namie town	50s female	House was completely flooded by the tsunami.
		50s female	Same as above
Fukushima Prefecture	Futaba town	60s male	Former Kisai High School refugee Tokyo Electric-affiliated company duty

Prefectures. The 24 victims dislocated by radiation emission due to the tsunami were from three cities and four towns including those whose entire towns were displaced. The foci of the interview questions were the practices of mutual help and community consciousness before and after the disasters at interviewees' original locations and then at communal shelters. The responses were divided into distinctive meaning units and analyzed through comparison according to categorization as purposeful samples by descriptive–interpretive qualitative research (Elliott & Timulak, 2005; Flick, 2014; Rubin, 2005). The methodological goal was to create a subset (the sample) from a larger set (the population) that was representative of the voices of the victims as a whole.

The first set of interviews were conducted about 1 month after the disaster (April 2011) surveying the following items: the condition of shelter life (from the moment of displacement up to the interview) and mutual help in regional societies before and after the disaster (interviewees' recognition of traditional mutual help as reciprocity, redistribution, and unidirection and the relation between traditional mutual help and the disaster).[2] In the interviews with different refugees conducted 8 months and 1 year after the disaster (November 2011 and March 2012, respectively), the survey focused on the conditions of shelter life (the changes and similarities from the beginning of displacement until the interview), their experience of mutual help (bonds among the victims and connection to outside residents), public regional planning, and possibilities of restoring communities. The interviews were conducted in accordance with ethical standards.

RESULTS

Community Consciousness of the Victims of the Earthquake and Tsunami

The Victims Who Are Conscious of the Bonds of Regional Societies

There were both strong and weak bonds in regional societies before the disaster. Residents who had lived in apartments or housing complexes may have had weaker ties, and so tended to be more conscious of the newly developed bonds among evacuees acknowledging the kindness of volunteers. A man in his 50s living in an apartment in Kitaibaraki City, Ibaraki Prefecture, said that he became acquainted with local victims after the disaster and was content with the life of the refugee because officials and volunteers helped him and communicated with him. A man in his 70s who had not been familiar with new neighbors was conscious of the "refugee community" because he could communicate with the victims, although they were from outside of his regional society. A daughter caring for her mother in her 80s in the same city said that, although she and her parent did not lose everything, people who lost something experienced the need for mutual help acutely. Her mother, however, said that there had been traditional mutual help networks, but these have disappeared. Yet she acknowledged that she and others shared food and supplies in the shelter through communal help and felt a bond of victimhood.

A man in his 70s in Ishinomaki City, Miyagi Prefecture, expressed despair that local people had no vision for the future, and it was more terrible than after the end of the Second World War when they still had hopes and dreams for the future. Those villages that had already been undergoing depopulation and an aging of their society were struck especially hard by the event. He belonged to a regional organization of 150 households that had a savings fund. The fund was distributed among members after the disaster. Although many residents left the society, some people who stayed in the district hope that their organization can be restarted to remake the community. Although the sentiments he expressed were negative, this is also one piece of evidence that mutual help in regional activity has not disappeared after the disaster.

Victims Who Are Skeptical of the Bond of Regional Societies

However, many respondents spoke of the breaking down of community bonds. A woman in her 50s said that the relationship among people in Rikuzentakata City, Iwate Prefecture, has been altered by the disaster and the community had been broken. Mass media covered victims living in public shelters but ignored other victims who could flee into private homes. As conditions in public shelters improved, those in private shelters demanded better care as well. There was a contradiction among the attitude of the more seriously afflicted toward inside and outside support. On the one hand, little feeling of gratitude was expressed toward other, less afflicted victims who offered assistance. Alternately, those seriously afflicted did express gratitude to official and volunteer aid. This difference clearly demonstrates a dissolution of community cohesion.

Community Consciousness in the Face of Nuclear Radiation

Victims Who Hope Their Communities Will Be Restored

As the local economy in evacuated towns had depended directly and indirectly on the nuclear power plant, some interviewees showed faith in their communities. A woman in her 30s from Okuma Town said that the escape from the disaster united residents, and she felt solidarity when she heard they should go to the town office by email. She added that residents who worked at Tokyo Electric Power Company (TEPCO) subsidiaries did not complain about the company and supported it to manage the accident. This suggests that there is a certain community consciousness among the residents as a "nuclear power town." Furthermore, she spoke of having been prepared for a nuclear plant accident including having been taught by her mother that an emergency iodine supply was available in the local government office.

A man in his 60s who fled from the disaster with a group from Futaba Town expressed conflicting feelings. On one hand, he definitely held TEPCO responsible for the accident. On the other hand, he still hoped that TEPCO would, in fact, reconstruct the town. A man in his 40s who had been engaged in car repair had customers who worked in the plant and its affiliates. He said that local people through neighborhood associations helped each other. He did not criticize TEPCO for the accident, because the company had contributed in the development of the town and his business. Although he knew few people in the shelter, he said that he was developing close feelings for some people with whom he lived in the same room.

People who work on cleaning up the accident are viewed in a favorable light, because many local people lived in the towns for the sake of employment at the Fukushima nuclear power plant and its affiliates. Before the construction of the plant, local farmers had looked for work outside of the town in the off season. The new plant meant that people could then work in local affiliated companies, leading to the image of "nuclear power towns" as places of prosperity.

Victims Who Feel It Will Be Difficult to Restore Communities
However, other voices show that the accident inflicted great damage on their societies. People who were displaced by the nuclear accident express deeper distress and anger than those who suffered "only" the natural disasters. Interviewees who were evacuees from towns affected by the nuclear disaster can be divided into individuals who fled the town individually and those who were evacuated in groups by the authorities. This is a useful distinction to make because their views of the community are affected accordingly.

Individual Evacuees. The individual evacuees are victims who did not flee in the town groups. The livelihoods of local producers including both farmers and fishermen were severely damaged by the triple catastrophe. A farmer in his 50s from Minamisouma City said that residents were left completely uninformed about the power plants after the tsunami. They endure the "invisible fear" of the radioactive accident. After the farmer came back to his town with his wife, bulldozers cleaned most debris off his land, but sand from the tsunami remained as well as glass and metal fragments that had hindered his rice crop. Furthermore, the farmer was intensely angry, because he could not bring up a meter for the rice restrictions imposed by the government. "We are at loss because we do not know how to cope with the accident and make a claim against the big company supported by the government." He decided to leave his land and find new land to farm.

A woman in her 50s who had lived in Namie Town said that the residential bonds had dissolved and her town would become "a ghost town." A man in his 60s from Minamisouma City said, "Though we could contact neighbors by telephone from the refuge, it was very difficult for us to have solidarity as regional residents because we were separated to flee. Local people from the city were helpful, but after the disaster they were isolated." He said that many victims were in a complete state of anxiety about the future and added that most people who wanted to go back home were elderly, but younger people, those under 40, would not be going back.

A woman in her 50s who had lived in Namie Town and whose house was washed away by the tsunami said that though her local society used to demonstrate solidary through cooperative ditch cleaning and her town previously had been connected with the neighboring town for funeral ceremonies, their relationships were disconnected after the disaster. A man in his 20s from Tomioka Town said, "We used to feel closer to our neighbors," and quoted the common Japanese saying, "A good neighbor is better than a brother far off." However, now he feels differently, amending the idea to "A stranger from a far is better than a relative nearby," as a way to express his gratitude to volunteers, while complaining about the actions of his local government.

Group Evacuees. People who fled in the Futaba Town groups feel the difficulty of maintaining solidarity in spite of the group character of the shelter population. About 1,400 people who were evacuated from the area near the Fukushima nuclear power plant first went to Saitama City 210 km away from the town and then ended up in a disused high school in Kazo City in Saitama Prefecture.[3] They continue to live as evacuees. The following interviews were conducted with the victims in April after the disaster. The comments of the interviewees exposed at least frustration and some were highly critical.

A man in his 60s said that his local activities had been vigorous before the disaster, being engaged in "the new life movement" and keeping a suitable life, and local people had previously lived together obeying communal rules. He emphasized the poor quality of human relationships resulting from evacuee life, because people were assigned to accommodations without regard to previous affinities as if in "a combined military unit." Another man in his 60s said that he participated in the "union of the dead (*Shibito*)," consisting of 17 households that helped during funeral ceremonies. When he started to live in the gymnasium, victims were separated into groups that formed a kind of residential unit, but he does not always feel that he has much of a relationship with any of them.

A woman in her 70s said that her regional society had neighborhood associations consisting of 12 households. Members saved money to pay 10,000 yen each year. The group used the money to travel or pay for a year-end party and an annual convention. However, after months as evacuees, they decided to dissolve the organization and distribute the money. She maintained that there had been a strong regional bond in the town. Furthermore, she said that her father had been almost completely ostracized because he had refused to sell his land to TEPCO to build the nuclear plant, but many farmers, lacking education, decided to sell out for what seemed like a lot of money without considering the consequences of living with a nuclear plant in the neighborhood. She has been living in the shelter with other members trying to maintain some semblance of collective solidarity. A man in his 60s insisted that the roadmap for going back home should be planned by the local government and said that it was difficult for the victims to unite as community although they had a vague feeling as a "refugee community" simply from obeying the rules in group living.

The following interviews were done 8 months after the disasters struck. A woman in her 30s said that she was learning what kind of residents lived in

Futaba Town. She described how these people were "losing their common sense." She also added that so far as she felt uneasiness, she could not go even one step forward. She became sick because of group life and her uncertainty about the future after she evacuated to the refugee shelter. Earthquake disaster-related deaths occurred because of poor health, overwork, and no bonds resulting from refugee life. A man in his 60s also expressed anxiety, adding that people could no longer help each other through neighborhood associations because of separation due to evacuation.

DISCUSSION

Possibility for Restoration of Communities

Transformation of Mutual Help Networks Inside and Outside of Regional Societies

Community life means that local people trust each other and abide by norms through stable mutual help networks as a spontaneous social order. Generally speaking, if mutual help networks are weak, a community needs outside help. If they are strong, it does not need outside assistance. Typically if weak internal networks do not receive support, they become weaker and weaker. The networks become stronger and stronger, if they receive such help. Contrarily, mutual help networks may be weakened sometimes if a community accepts outside help that ends up supplanting them. The relation between inside and outside networks is not substitutional but complemental. In light of this observation in the interview data, three propositions are made.

The First Proposition. Those victims who felt that bonds in a regional society were thin and whose connections with others were not strong congratulated the help of volunteers and committed themselves to the "refugee community." The first proposition about the transformation of mutual help networks, then, is that the less intimate human relations in a regional society are, the more intimately the victims respond to the help of volunteers (Fig. 4). Those who had been living in apartments or housing complexes seemed content with their "shelter community."

The Second Proposition. A woman in her 50s from Rikuzentakata City, Iwate Prefecture, said, "The disaster broke human relations completely. New relations developed between people whose houses were swept away by the tsunami and those who experienced significantly less damage." Those whose

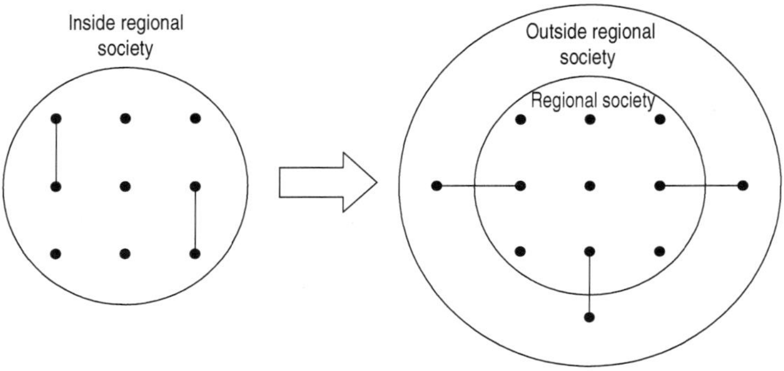

Proposition 1: Mutual help networks connecting outside regional society

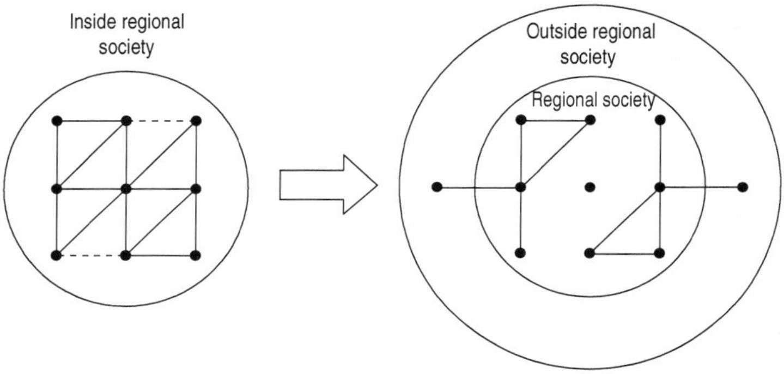

Proposition 2: Mutual help networks being divided in regional society

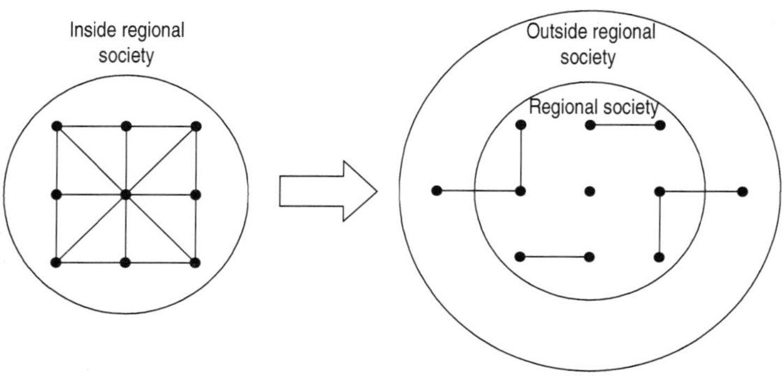

Proposition 3: Mutual help networks being cut in regional society

Fig. 4. Transformation of Mutual Help Networks.

homes were not completely destroyed gave room and food to those whose homes had disappeared. However, the seriously afflicted victims felt that the difference between the life of the public shelter, where other victims were fully helped by the local government and volunteers, and the life in a private refuge like habitable houses must be evident and that public shelter life was much better. The more those whose houses were swept away felt discriminated against, the bigger the gap among local people grew. There was an inner voice: "What help on the earth do you give us?" "You" means people whose houses had not been disappeared. Those seriously afflicted were thankful for the outside volunteers. Some of less afflicted victims thought that it was better not to help them at all. The woman had heard that there were the same voices in the city.

It might be assumed that bonds generally became strong through the disaster. However, as this case shows, the social bonds among both groups of victims have changed. Those whose houses were not destroyed by the tsunami have hesitated to communicate with less fortunate neighbors because of "survivors guilt." Thus, a second proposition about the transformation of mutual help networks can be drawn from this evidence. It is that the stronger the outside help, the more the victims rely on it and tend to ignore local help, weakening local bonds (Fig. 4). As a result, local society is further fractured. There is also unchangeable relationship among local people whom the disaster assailed in other areas. There might be no friction of feelings among victims who share the fear of radioactive contamination.

The Third Proposition. A woman in her 50s from Futaba Town living in the refuge said, "We discussed an organization of parents for school, but it was not realized because we are separated." The "refugee community" does not really function now. The bonds have been cut among not only victims in the shelter but also residents in the community. Many refugees are afraid that the town will never reappear. The longer they live as refugees, the worse human relations among them become. A woman in her 30s said, "The more closely we live, the more clearly we can see each other's previously unnoticed bad habits. We cannot say what we want to say in front of roommates. I have come to read their faces. Some people say the bonds are strengthening, but I do not believe it. There are no bonds here at all." Many other victims echo these observations. A man in his 60s said that he expected to find the persuasive power of mutual help. The "shelter community" would rediscover previous bonds, he thought, but it had not happened. After feeling this disappointment, members began to feel themselves to be part of a "refugee quasi-community."

The third proposition is that if the lack of mutual help networks in a separated society dissolves previous bonds, then local people will look to recreate them elsewhere (Fig. 4). Traditional mutual help networks as a spontaneous social order has played an important role for connecting local people in Japan and elsewhere (Onda, 2006, 2013). A community with a low degree of horizontal integration has a weakly knit social network among local people (Berke, Kartez, & Wenger, 1993). The more closely they had previously acted through traditional mutual help networks, the more they were disappointed with the loss of these relationships.

Outside Support and Inside Support

The limits of victim support in light of the three stated propositions should be recognized. At the same time, Japanese policy makers and disaster volunteers have to listen to the victims' heartfelt cries. It is important for them to be mindful of the mentality of local people as well as listen for the unvoiced sorrows of the socially vulnerable. The needs of those assisted and the role of those assisting should be understood (Quarantelli, 1999). Volunteers should see not the mud but the human beings when they help victims. The ability to recognize and respond to the victims' unspoken needs as well as their sometimes unspoken gratitude is essential for healing to occur. Outside support, inside support among local people, and self-support should be distinguished in restoring communities.

Mutual Help in Regional Societies and Restoration of Communities

Possibility of Restoration of Communities Following the Disaster Seen from the Community Life Cycle

Considering community consciousness, a life cycle divided into four phases can be constructed. The first is a "period of introduction" in which community consciousness starts to grow when new residents begin to live as neighbors. This is followed by a "period of growth." The third phase is a "period of maturity" during which a community heads toward stasis and starts to scatter. The final phase is a "period of decline" when growth is completely curtailed and population dwindles. If a community is in the fourth stage, means of revitalization need to be considered. By what means could it be possible to revive community consciousness?

Participation is important for building community (Abbott, 1995). Taking the roots of the word community literally, "com" meaning to be communal and "unis" to have responsibility, it can be inferred that local people have to rebuild the community through recreating communal responsibility (MacIver, 1917). On the basis of the first proposition developed in the study through the experience of the victims, then reconstruction of a local disaster prevention organization should be done through participation by neighbors whose previous community ties were weak.

As the second proposition shows, outside volunteers can gain the confidence of local people, acting as "catalyst" that could lead to repairing the cracks in devastated community networks. This can be effective even though, in the end, the previous relationships are unlikely to be perfectly reconstituted. Outside help should first play the role of supplementing mutual help networks in local societies. The best abilities of networkers and coordinators of volunteers facilitate a codependence of healing leading to eventual independence of the outside support systems. Mutual help among local people is important because they know better the psychological or emotional needs of the others in the community.

However, it is much more difficult for the victims of the added nuclear disaster to restore their communities. Many people were compelled to leave their hometowns and to flee Fukushima Prefecture entirely. However, the third proposition shows it to be important for them to rebuild their communities. In light of what is known about community life cycles, local peoples' community consciousness in their new settled locations should be reconsidered. The reality is that they have to resettle outside of the contaminated area. Some local units of the residents have been assembling to meet each other and to hold tea parties similar to the previous periodical meetings of the area. The smaller the townships are, the more intimate and cohesive the local people are. New nodule organizations where new and old inhabitants can interact through small group activities that can decrease the newcomers' uncertainty and loneliness need to be established. It is more preferable that the residential units of people who lived in the same town previously should be reassembled to replant the old communities more successfully. Structural cohesion as social relationship embedded in a regional society needs to be considerable (Moody & White, 2003).

Volunteers might play a role in connecting the victims who have become scattered as "weak ties" or connecting scattered communities as "structural holes" (Burt, 1992; Crow, 2004; Granovetter, 1975). If local people in the same town can assume this role, the bonds of victims and the cohesiveness of the communities would be strong (Hechter, 1987; Olson, 1965). After all, it could be recommended that victims restore mutual trust and rebuild community empowerment again by themselves. Recovery is not putting things back the way they were before, but means adapting to new realities (Alesch, Arendt, & Holly, 2009). However, the recovery aims at not only improving individuals' capabilities (Gardoni & Murphy, 2010) but also getting community empowerment. The local government has been trying to continue keeping residents' community consciousness.[4]

Lessons for Restoring Communities Through the Big Disaster

"Forced depopulation of the stricken area" has masked the severe problem of falling economic vitality due to depopulation that had been naturally occurring in much of rural Japan before the disaster. Especially, more intensive investment by local governments is necessary for the recovery of merged communities. The less populated an area becomes, the less vital the area will be. It is estimated that the population decreased by approximately 72,000 people in 40 suffering areas between March 1, 2011, and February 1, 2013. The idea that the reconstruction of some areas is not necessary simply because the victims do not come back should not be embraced.

The problem of the radiation-contaminated areas is very difficult to solve, because the victims may never be able to go back to their hometowns. Many victims who had fled felt that the town was not united and the community was broken up. A woman in her 30s said that although the town mayor would make a new town outside their former town, it would not be the same town. A man in his 60s said that elderly people wanted to go back, but it was only nostalgia. A sustainable environment should be provided by national and local governments and conditions improved for current and future generations (Lizarralde, Johnson, & Davidson, 2009). The younger the evacuees are, the less intention they have to go back to their hometowns. Young couples are afraid of the effect of possible radiation exposure on their children. Some local governments are on the brink of extinction because of population decline. As a result, a merger of associated local governments is being considered.

In one case, an election for town mayor was contested over the main issue of the transfer of the entire town to a new location. The candidate who had insisted on returning to the original town site and rebuilding there won. Nonetheless, the road to recovery is difficult for the local governments that were forced into evacuation. However, building a contemporary town, meaning physical rebuilding in other places near the original towns, is realistic for the evacuees whose hometowns cannot be reoccupied in the near future.[5] There are two ways of building contemporary towns being planned by the communities of Namie, Futaba, Okuma, and Tomioka. The evacuees of the four towns number about 40,000 of the total of 150,000 in Fukushima Prefecture as a whole. One possibility is to build an entire new town including homes, schools, and hospitals in a new location and transfer the displaced populations to the new town site. In contrast with this "integrated type," the other choice is a "dispersed type," building apartment complexes consisting of some houses each in the cities of Koriyama, Iwaki, and Aizuwakamatsu in Fukushima. The new arrived citizens would use the existing infrastructure for services.

Local governments that approve the first solution are afraid that the newly built contemporary towns might become "ghost towns," empty of people if the evacuees eventually do have the option to return to a decontaminated original hometown sometime in the future. Furthermore, the national and prefectural governments are also hesitant to back this plan, because there has been no precedent to rely on as a predictor of success. Japanese officials tend to be extremely cautious about such costly, complex projects. Therefore, the dispersed resettlement option is coming to be seen as the most realistic. It has been argued that planners and policy makers put local people before plans (Ganapati, 2009). The transfer authorities should note that the units of transference should be considered to be separated according to former residential communities so that traditional mutual help networks could more easily maintain their vitality.

On the second anniversary of the disaster, a staff member of Futaba Town office in his 60s said that mutual help networks had weakened after "mechanization," referring to a time at the beginning of regional development when some communities changed to become nuclear plant service towns, enabling farmers to stop working away from home in the slack season. He meant that society should reevaluate the cost of development, both then and now. In the case that the community itself is destroyed, social capital cannot function (Coleman, 1990). It presumes the continuance of community. However, social capital as trust and norms of reciprocity is often maintained in recovering community after disaster (Bridger & Alter, 2006; Forrest & Kearns, 2001; Woolcock & Narayan, 2000). The recovery of mutual help networks among local people as a spontaneous social order is important.

One final problem is that the number of elderly people is increasing in the stricken area. Many of them have become weakened physically and got sick after the disaster, because they do not venture out from their temporary housing and are unfamiliar with the new life. Some specialized nursing homes for them were damaged and closed, reducing options for care. There is a worry that such people will suffer isolated illnesses and deaths resulting from their continuing highly stressful circumstances. Putting people at the center of disaster and development should be emphasized (Collins, 2009). Rebuilding community is not reformation but restoration (Birch & Wachter, 2006).

Post-disaster housing reconstruction must avoid rebuilding structures that reflect, sustain, and reproduce social inequality, and participation of local people is important for successful resettlement (Oliver-Smith, 1991). It might be difficult for the newcomers to continue social and cultural conditions in their new communities. The lack of cultural continuity and compatibility results in increasing disaster vulnerability after post-disaster reconstruction. Community empowerment as resilience or resiliency is important. Communities with robust social networks, with an abundance of social capital, are better able to coordinate recovery and to minimize the migration of people out of the area. Local governments should listen to local peoples' evaluation of the success of the recovery (Kweit, 2004).

CONCLUSIONS

If a town were physically rebuilt, would the community mind rebuild as well? It is necessary for the victims to recreate and maintain not "an imitation community" in the shelter but "a genuine community." Evidence provided by the voices of the victims together with the three propositions for the transformation of mutual help networks lead to these conclusions. The proposition that a disaster brings neighbors closer to each other does not necessarily hold in the face of truly devastating loss. The stronger the previous ties of the victims were, the more fragile they felt when the bonds were materially fractured creating a wall of separation between those whose houses were left standing and those who lost everything. Inside communal help became weaker and weaker and the outside assistance, from volunteers whom the victims came to trust and rely on, became stronger and stronger. Mutual trust makes collective understanding easier (Putnam, 1993). It is important for the victims to help each other, because they know the pain of the same circumstances and the degree of sympathy is not different from that offered by outsiders. In some cases, outsiders' intervention as network facilitators might be necessary to the restoration of communal bonds, acting as catalysts to connect widely scattered evacuees.

This study has shown that an unprecedented disaster overcame a regional society not only physically but also devastated local morale, and it is important to restore mutual help networks as a spontaneous social order. The damage to the victims who suffered both the natural and nuclear disasters is greater than that of those who suffered from "only" the earthquake and tsunami because of confronting the greater loss of regional society itself. Anxiety about the future seems to have served to reduce individual connectedness and further crack communal bonds. Many people abandoned the idea of return to their previous homes and decided to permanently relocate. Therefore, it is very difficult to reunite as community that local people had formed and occupied all their lives. New communities should be constructed from and with as many former neighbors and residents as possible. Responsible officials should especially focus on the least advantaged members of regional society and the socially vulnerable elderly who hope to return to their former locations (Ingram, Franco, Rio, & Khazai, 2006).

Timely and rapid public help is necessary to restore and revive regional societies physically and institutionally. However, reconstruction encompasses not only the recovery of visible indicators (Beniya, 2007; Haas, Kates, & Bowden, 1977) but also that of community consciousness. A sustainable community made up of a functioning social sphere would not rely solely on outside help as public assistance but instead on the responsibility that local people shoulder (Monday, 2001). Communal help among local people is indispensable to regain relationships and bonds in communities, and strong intention to improve each other's situations through self-help is also necessary. The trinity of public help, communal help, and self-help is important for the restoration of the communities (Onda, 2006, 2013). We have been confronting reality to restore and revive not only the stricken areas but also the super-aging society with a declining population. As Tolstoy wrote in Anna Karenina: "Happy families are all alike; every unhappy family is unhappy in its own way." We Japanese should put this difficult time behind us but advance toward the future not forgetting its historical lessons.

NOTES

1. Total deaths from the combined disasters number 18,486 including 2,597 missing persons (data as of November 10, 2014, according to the National Police Agency) and the number of the evacuees were approximately 470,000 (March 2011 according to the National Reconstruction Agency).

2. Japanese mutual help actions are divided into three categories (Onda, 2006, 2013). One is reciprocity in helping to plant rice and re-roof houses by exchanging labor. The second is redistribution. In exchange for the right to get goods from a common store, local people have the obligation to maintain a common pool of resources. Finally, unidirection refers to support in funeral and wedding ceremonies requiring no monetary exchange.

3. All victims displaced by radiation had left the school by December in 2013. The shelter as a "refugee community" was closed at the end of March in 2014.

4. The local government of Futaba Town has been distributing public information magazines periodically not only to each household but also to evacuees who have been living apart from their extended families, where addresses are known. This is one means the government is using in its hope to maintain a sense of unity among the townspeople. Although some evacuees transferred their resident registration cards, signaling permanent relocation to new places, the magazines have been distributed to them as well, as Futaba residents on March 11, 2011, when the disaster occurred. In addition, a public office work association consisting of six towns and two villages near the Fukushima nuclear power plant have installed live cameras to enable evacuees with PC access to see live pictures displayed on the towns' Internet home pages.

5. In 2012, the central government designated three zones of nuclear contamination: the first area is difficult to go back to the hometown, the second area is restricted for current residence but possible to return within 5 years following the disaster, and the third area is possible for return pending cancellation of the evacuation directions at some unspecified earlier time. The areas assessed comprise the 11 towns nearest the Fukushima nuclear power plant.

ACKNOWLEDGMENTS

I would like to thank my faculty colleague Professor Robert B. Spenser and Adam Gyenes for their helpful comments and suggestions.

REFERENCES

Abbott, J. (1995). Community participation and its relationship to community development. *Community Development Journal*, 30, 158–168.

Aldrich, D. (2012) *Building resilience: Social capital in post-disaster recovery*. Chicago, IL: University of Chicago Press.

Alesch, D. J., Arendt, L. A., & Holly, J. N. (2009). *Managing for long-term community recovery in the aftermath of disaster*. Fairfax, VA: Public Entity Risk Institute.

Beniya, S. (2007). The evaluation of the status of disaster areas by using recovery indicators (in the case of the Great Hanshin-Awaji Earthquake). In *Proceedings of the 2nd International Conference on Urban Disaster Reduction*, Taipei, Taiwan, November 27–29.

Berke, P. R., & Campanella, T. J. (2006). Planning for postdisaster resiliency. *The ANNALS of the American Academy of Political and Social Science*, *604*, 192–207.

Berke, P. R, Kartez, J., & Wenger, D. (1993). Recovery after disaster: Achieving sustainable development, mitigation and equity. *Disasters*, *17*(2), 93–109.

Birch, E. L., & Wachter, S.M. (2006). *Rebuilding urban places after disaster: Lessons from Hurricane Katrina*. Philadelphia, PA: University of Pennsylvania Press.

Boen, T., & Jigyasu, R. (2005). Cultural considerations for post disaster reconstruction post-tsunami challenges. In *UNDP Conference* Kobe, Japan, January 18–22 (pp. 1–10).

Bourque, L. B., Shoaf, K. I., & Nguyen, L. H. (1997). Survey research. *International Journal of Mass Emergencies and Disasters*, *15*(1), 71–101.

Bridger, J., & Alter, T. (2006). Place, community development, and social capital. *Community Development*, *37*, 5–18.

Burt, R. S. (1992). *Structural holes.* Cambridge, MA: Harvard University Press.

Chang, S.E. (2010). Urban disaster recovery: A measurement framework and its application to the 1995 Kobe earthquake. *Disasters*, *34*(2), 303–327.

Coleman, J. S. (1990). *Foundation of social theory*. Cambridge: Harvard University Press.

Collins, A. E. (2009). *Disaster and development*. London, UK: Routledge.

Crow, G. (2004). Social networks and social exclusion. In Phillipson, C., Allan, G., & Morgan, D. (Eds.), *Social networks and social exclusion* (pp. 7–19). Aldershot, UK: Ashgate.

Elliott, R. & Timulak, L. (2005). Descriptive and interpretive approaches to qualitative research. In: Miles, J. & Gilbert, P. (Eds.), *A handbook of research methods for clinical and health psychology* (pp. 147–159). Oxford, UK: Oxford University Press.

Flick, U. (2014). *The SAGE handbook of qualitative data analysis*. Los Angeles, CA: SAGE Publications.

Forrest, R. & Kearns, A. (2001). Social cohesion, social capital and the neighborhood. *Urban Studies*, 38, 2125–2143.

Ganapati, N. E., & Ganapati, S. (2009). Enabling participatory planning after disasters. *Journal of the American Planning Association*, 75, 41–59.

Gardoni, P., & Murphy, C. (2008). Recovery from natural and man-made disasters as capabilities restoration and enhancement. *International Journal of Sustainable Development Planning*, *3*(4), 1–17.

Gardoni, P., & Murphy, C. (2010). Gauging the societal impacts of natural disasters using a capability approach. *Disasters*, 34, 619–636.

Granovetter, M. S. (1975). The strength of weak ties. *American Journal of Sociology*, 78, 1360–1380.

Haas, J. E., Kates, R. W., & Bowden, M. J. (1977). *Reconstruction following disaster*. Cambridge, MA: MIT Press.

Hawkins, R. L. & Maurer, K. (2009). Bonding, bridging and linking: How social capital operated in New Orleans following Hurricane Katrina. *British Journal of Social Work*, *40*, 1777–1793.

Hechter, M. (1987). *Principles of group solidarity*. Oakland, CA: University of California Press.

Ingram, J. C, Franco, G., Rio, C. R., & Khazai, B. (2006). Post-disaster recovery dilemmas: Challenges in balancing short-term and long-term needs for vulnerability reduction. *Environmental Science & Policy*, *9*(7), 607–613.

Kweit, M. G., & Kweit R.W. (2004). Citizen participation and citizen evaluation in disaster recovery. *The American Review of Public Administration*, *34*, 354.

Lizarralde, G., Johnson, C., & Davidson, C. (Eds.). (2009). *Rebuilding after disasters: From emergency to sustainability.* New York, NY: Spon Press.

MacIver, R.M. (1917) [1924]. *Community: A sociological study*. London, UK: Macmillan.

Mochizuki, J. (2014). Decision-making, policy choices and community rebuilding after the Tohoku Disaster. *Journal of Integrated Disaster Risk Management*, *4*(2), 103–118.

Monday, J. (2001). Introduction to sustainability. *Holistic disaster recovery: Ideas for building local sustainability after a natural disaster* (pp. 1–11). Boulder, CO: National Hazards Research and Applications Information Center.

Moody, J. A., & White, D. R. (2003). Structural cohesion and embeddedness: A hierarchical concept of social groups. *American Sociological Review*, 68, 103–127.
Murakami, K., Murakami, W. D., Tomita, H., Miyake, S., Shiraki, R., Itonaga, K., & Dimmer, C. (2014). Planning innovation and post-disaster reconstruction: The case of Tohoku, Japan/reconstruction of tsunami-devastated fishing villages in the Tohoku region of Japan and the challenges for planning/post-disaster reconstruction in Iwate and new planning challenges for Japan/towards a "network community" for the displaced town of Namie, Fukushima/resilience design and community support in Iitate Village in the aftermath of the Fukushima Daiichi nuclear disaster/evolving place governance innovations and pluralising reconstruction practices in post-disaster Japan. *Planning Theory & Practice*, 15(2), 237–265.
Nakagawa, Y. & Shaw, R. (2004). Social capital: A missing link to disaster recovery. *International Journal of Mass Emergencies and Disasters*, *22*(1), 5–34.
Oliver-Smith, A. (1991). Successes and failures in post-disaster resettlement. *Disasters*, *15*, 12–23.
Olson, M. (1965). *The logic of collective action: Public goods and the theory of groups*. Harvard, MA: Harvard University Press.
Onda, M. (2006). *Gojo Syakai Ron* (Mutual Help Society) [in Japanese]. Kyoto, JP: Sekaishisousya.
Onda, M. (2013). Mutual help networks and social transformation in Japan. *American Journal of Economics and Sociology*, *71*(3), 531–564.
Putnam, R. D. (1993). *Making democracy work*. Princeton, NJ: Princeton University Press.
Quarantelli, E.L. (1999). *The disaster recovery process: What we know and do not know from research*. Newark, DE: University of Delaware.
Rodríguez, H., Quarantelli, E.L., & Dynes, R.R. (2006). *Handbook of disaster research*. New York, NY: Springer.
Rubin, H. J., & Rubin, I. S. (2005). *Qualitative interviewing: The art of hearing data*. Thousand Oaks, CA: SAGE Publications.
Tatsuki, S. (2007). Long-term life recovery processes among survivors of the 1995 Kobe Earthquake: 1999, 2001, 2003, and 2005 life recovery social survey results. *Journal of Disaster Research*, *2*(6), 484–500.
Wassermann, S., & Faust, K. (2009). Social network analysis: Methods and applications. New York, NY: Cambridge University Press.
Woolcock, M. & Narayan, D. (2000). Social capital: Implications for development theory, research and policy. *The World Bank Research Observer*, 18, 293–249.

CHAPTER 5

LESSONS FROM DISASTER RECOVERY IN JAPAN THROUGH CASE STUDIES OF FOUR EARTHQUAKES

Yingying Sun

ABSTRACT

This chapter focuses on the experiences and processes of earthquake recovery to discuss the general state of disaster recovery in Japan. In this way, it is expected that the outcome and discussion can provide effective insights for both domestic and international disaster-related efforts. First, the characteristics of natural disasters in Japan are summarized using statistical data. Quantitative and qualitative methods are flexibly used to analyze published data, materials, and semistructured interview data. Published data and materials are collected from various sources. Interview data were gained from diverse interviewees. Then, four case studies of earthquake recovery are introduced and the application of their recovery experiences to future disaster risk reduction is proposed. Finally, conclusions have been drawn from these case studies to show the practical influence of disaster-recovery experiences to regions that are currently experiencing or are likely to experience natural disasters in the future. More specifically, the chapter

Recovering from Catastrophic Disaster in Asia
Community, Environment and Disaster Risk Management, Volume 18, 85–104

ISSN: 2040-7262/doi:10.1108/S2040-726220160000018004

illustrates what challenges and influences past earthquakes can have on our present preparedness against a Nankai Trough Earthquake, which is predicted to occur in the near future.

Keywords: disaster recovery; urban earthquake; rural earthquake; civilian groups; disaster volunteer center; Nankai Trough Earthquake

INTRODUCTION

Natural disasters in Japan comprise earthquakes, tsunamis, volcanic eruptions, meteorological and climate disasters, as well as others (e.g., landslides and technological disasters secondarily triggered by natural disasters). As a country frequently affected by natural disasters, Japan has established several systems of political, organizational, and civil countermeasures to cope with their consequences. As a result, a large volume of diverse disaster recovery experiences has been accumulated, some of which have been successfully applied in the recovery processes of later disasters or in disaster risk reduction (DRR) measures for potential future disasters.

The contents of this chapter are as follows. First, the characteristics of natural disasters in Japan are summarized using statistical data. Then, after explanation of the methodologies, four case studies of earthquake recovery are introduced, and the application of their recovery experiences to future DRR is proposed. Finally, conclusions have been drawn from these case studies to show the practical influence of disaster-recovery experiences to regions that are currently experiencing or are likely to experience natural disasters in the future.

DISASTER CHARACTERISTICS OF JAPAN

In the past 50 years, Japan has experienced 39 major disasters, defined as flood or wind disasters causing more than 500 deaths, snow disasters causing more than 100 deaths, earthquakes causing more than 10 deaths, or any disaster where one or more Emergency Response Headquarters was established to cope with issues of rescue, reconstruction, and recovery (Statistics Japan, 2016). Together, these disasters have resulted in a total of 32,372 people missing or dead. Of the 39 disasters, 13 were earthquakes and earthquake-triggered tsunamis, which account for 89% of those reported missing or dead. The large death toll of earthquakes is a clear feature of disasters in Japan.

When the number of earthquakes in Japan is compared with the global total, the high frequency of earthquake occurrence in Japan stands out as another characteristic of Japanese disasters. According to the Japan Meteorological Agency's database (2016), during the period from 2003 to 2012, there was an average of 157 earthquakes measuring about 5.0 on the Richter scale per year. Data collected by the United States Geological Survey (2016) show that the average number of annual earthquakes globally for this same period was 1928, demonstrating that more than 8% of the world's earthquakes occurred in Japan. The lessons learned from the earthquake-recovery process in Japan may therefore have important indicative value for disaster recovery worldwide.

The 26 nonearthquake major disasters (about 67% of the 39 total) in the past 50 years comprised typhoons, volcanic eruptions, snow, and floods, accounting for 11% of the entire disaster death toll. The frequency of each disaster type, including all disasters instead of just major disasters, was lower than that of earthquakes (157 per year); for example, there were only 22 typhoons per year for 2003–2012 (Japan Meteorological Agency, 2016). Therefore, disaster-recovery methods in Japan can be represented primarily by earthquake-recovery methods.

The aforementioned situation of disaster characteristics in Japan shows that disaster recovery and DRR for earthquakes is a clear priority. As a result, this chapter focuses on the experiences and processes of earthquake recovery to discuss the general state of disaster recovery in Japan. In this way, we hope the outcome and discussion of this chapter can provide effective insights for both domestic and international disaster-related efforts.

METHODOLOGIES

Four case studies of the processes and experiences of earthquake recovery, and their application to DRR, are introduced in this chapter. Quantitative and qualitative methods are flexibly used to analyze published data, materials, and semistructured interview data. Published data and materials are collected from various sources, such as reports from civic agencies, research institutes, and database banks. Interview data were gained from diverse interviewees, such as government officials of the Emergency Management Department, earthquake victims, and community members living in earthquake-prone areas.

LITERATURE REVIEW

Disaster recovery has been extensively discussed by various stakeholders, such as community members, national and local governments, mass media, and academic scholars. The analytical focuses of each group vary due to differences in their concerns and objectives of discussion; however, they can be broadly categorized into studies for practical purposes and those for academic purposes.

Practical discussions of disaster recovery were usually seen as part of verification projects conducted by local governments in disaster-affected areas. Important items assessed in these projects regarded the population, economy, residential condition and community development, financial situation, citizen happiness index, disaster memorial museums, and integrated disaster education and research (Kobe City, 2014; Niigata Prefecture, 2015). For example, the item on population included information about citizens' lives, physical and mental health, citizen autonomy and mutual aid, social welfare for the elderly and those with disabilities, children's development, lives of foreign citizens, and the promotion of volunteer organizations, nonprofit organizations (NPOs), and nongovernmental organization (NGOs).

The specific aspects of disaster recovery that are studied vary from researcher to researcher. Discussion of the roles and functions of community relationships in the process of recovery is the most common topic (Chamlee-Wright, 2011; Cretney, 2016; Paton, 2007). Capacity building and physical or mental health disorders or disabilities were another frequent research concern (Springgate, Wennerstrom, & Carriere 2011; Weissbecker, 2009). Quantitative studies aimed at analyzing the progress of reconstruction and recovery have also been seen in statistics, economics, and engineering.

Although these previous studies have yielded many achievements, research is still lacking in various areas such as systematic summaries and discussion of the disaster-recovery process in specific locations. One reason might be the flexible responses taken in response to different disasters, which prevents systematic discussion. Thus, by focusing on earthquake recovery in Japan, this chapter aims to provide an alternative direction for disaster-recovery studies.

CASE STUDIES

Although several earthquake disasters have occurred in Japan in the past several decades, this chapter focuses, in particular, on four earthquakes that greatly influenced disaster recovery and DRR policies. More specifically, the

Table 1. Summary of Five Earthquakes.

	Date	Magnitude	Deaths	Completely Destroyed Buildings
Great Hanshin Earthquake	January 17, 1995	7.3	6,434	189,786
West Tottori Earthquake	October 6, 2000	7.3	0	395
Niigata Chuetsu Earthquake	October 23, 2004	6.8	68	3,175
Tohoku Earthquake and Tsunami	March 11, 2011	9.0	18,430	121,783
Nankai Trough Earthquake[a]	Unknown	>8.0	323,000	2,380,000

Note: [a]Values for the Nankai Trough Earthquake are based on projected values reported by the Cabinet Office (2012).

chapter illustrates what challenges and influences past earthquakes can have on our present preparedness against a Nankai Trough Earthquake, which is predicted to occur in the near future.

The policies and guidelines of disaster recovery in Japan are highly dependent on each local government agency, which results in diverse recovery experiences and achievements across different regions. This section introduces the processes of disaster recovery according to three categories: urban earthquakes, rural earthquakes, and widespread earthquakes possessing both urban and rural characteristics. Table 1 shows the basic information for each earthquake, and Fig. 1 shows their epicenters. The contents of each case study are presented as four components: (1) damage, (2) organizations and policies established or applied for recovery, (3) activities of civilian groups, and (4) contributions from community members.

Urban Earthquakes

An earthquake that occurs in a large municipality is usually recognized as an urban earthquake. The recovery processes for an urban earthquake could be much more complex than for a rural one due to the many different layers of population composition, industrial structures, and the relationships between civic agencies and the public. These harsh challenges were placed on the shoulders of Hyogo citizens in 1995.

Damage

The Great Hanshin Earthquake occurred on January 17, 1995, in the southern part of Hyogo Prefecture. It measured 7.3 on the moment magnitude

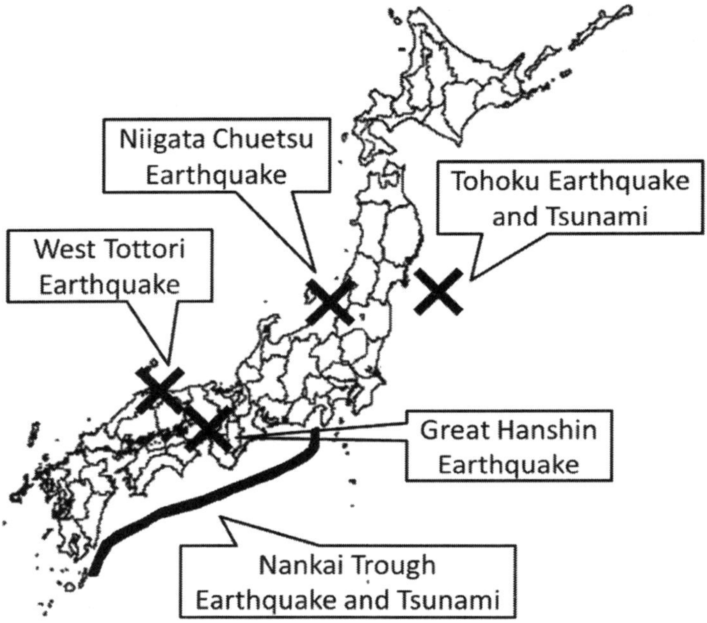

Fig. 1. Epicenters of Five Earthquakes.

scale. Up to 6,434 people lost their lives, with 189,786 houses completely destroyed, 7,500 of which are burnt down. The economic losses caused by the earthquake mounted up to 10 trillion JPY (Architectural Institute of Japan, 1995).

Organizations and Policies

After the earthquake, several organizations were established in the central and local governmental agencies. In hierarchical order, the central government established the Hanshin Recovery Committee, Hanshin Recovery Countermeasure Headquarters, and Association of Relevant Ministries of Hanshin Recovery (Hyogo Prefecture, 2016). At the local government level, five organizations were established: three aimed at promoting disaster countermeasures and two for recoveries. For example, the Integrated Department of Great Hanshin Earthquake Countermeasures included 13 deputy departments, which could be categorized into Emergency Countermeasure Departments, Reconstruction Departments, and Recovery Departments.

The function of the aforementioned organizations was to plan policies and countermeasures for creative reconstruction and recovery.

For example, a 3-year emergency recovery plan was made to rebuild infrastructure, accommodations, and industry aimed at restoring the area to pre-earthquake conditions. However, the Law of Natural Disaster Victims Relief was not enacted until 3 years after the earthquake (Hyogo Prefecture, 2016).

Activities of Civilian Groups

Owing to mass media reports of the seriousness of the Great Hanshin Earthquake, hundreds of thousands of people from outside Hyogo came and participated in volunteer activities. According to the verification report of volunteer activities at the 20th anniversary of the Great Hanshin Earthquake, a total of 1,380,000 volunteers assisted in rescue and recovery in the year 1995 (Hyogo Prefecture, 2016). Volunteer activities were diverse comprising efforts such as helping with shelter operation, carrying rescue materials, organizing events for community members, and providing mental health care. As a result, the year 1995 was recognized as the "first year of volunteers." Owing to the experiences and development of volunteer activities in the rescue and recovery process of the Great Hanshin Earthquake, volunteers became an indispensable part of efforts and attracted more and more participants.

However, because of the surprisingly large number of volunteers, problems emerged in volunteer logistics, such as unavailable housing for volunteers, inefficient coordination between volunteers and the needs of victims, and lack of any protocols for establishing volunteer centers. In this way, the Great Hanshin Earthquake revealed the need to prioritize coordination of volunteers and victims' needs.

Contributions from Community Members

During rescue efforts in the Great Hanshin Earthquake, 97.5% of victims pinned under collapsed buildings were freed by family members, community members, and passersby (Kawata, 1997). From the early stages of shelter life, victims supported each other. Volunteer activities were able to satisfy many local needs, victims voluntarily became shelter leaders, and many young people participated in volunteer activities.

When moved to temporary houses, victims started to organize temporary voluntary associations, enhance communications among residents, and run community centers. Also, NPOs and NGOs provided help to establish new communities and encourage communications between victims and other residents living nearby. However, because of the high percentage of aging residents in temporary houses, large gaps emerged in the ability to run voluntary associations in different communities. A sense of unfairness arose from the different level of support between victims who had begun to move out of temporary houses and those who had to move to temporary houses (Institute of Social Science, Waseda University, 1996).

Through these recovery experiences, the importance of mutual help and the neighborhood were recognized by the public at large. As a result, conventions on community development, *jishubo* (neighborhood associations for disaster preparedness) activities, and community businesses were implemented with consideration of creative recovery.

Rural Earthquakes

An earthquake that occurs in areas with low income, stagnated economic development, or depopulation is usually recognized as a rural earthquake. In Japan, a few such rural earthquakes have occurred in the past 50 years. To show the clear distinction from the Great Hanshin Earthquake, this section introduces two earthquakes: the West Tottori Earthquake and the Niigata Chuetsu Earthquake.

Damage

The West Tottori Earthquake occurred on October 6, 2000, with a magnitude of 7.3. Two months before this earthquake, the Tottori government conducted an earthquake drill, the results of which have been utilized in rescue activities. Several people pinned under collapsed buildings were rescued promptly, and the earthquake caused no loss of life. In addition, owing to the solid foundation of the ground and strong architectural mechanisms designed to cope with heavy snowfall, no more than 3,000 buildings completely or partially collapsed (Umeda, 2001).

The Niigata Chuetsu Earthquake occurred on October 23, 2004, with a magnitude of 6.8. Tens of hundreds of aftershocks were felt, some of which were extremely strong. The earthquake caused 68 deaths and completely or partially destroyed 18,000 buildings. Although there were many strong aftershocks, the number of destroyed buildings was limited, owing greatly to depopulation as well as to strong architectural mechanisms, as was the case in west Tottori (Niigata Prefecture, 2015).

Nonetheless, more than 6,000 locations were affected by landslides or avalanches due to a series of ten typhoons that had occurred before the earthquake, bringing heavy rains. As a result, many roads were shut down, and mountainous villages lost transportation lines and became cut off from the outside (Inagaki & Odagiri, 2014). For example, the entire population of Yamakoshi Village was evacuated by Japan Self Defense Forces helicopters. Some tunnels on main roads also collapsed, preventing the transportation of emergency goods.

Organizations and Policies

An Emergency Response Headquarter was established immediately after the West Tottori Earthquake, with the governor of Tottori as its head. At the same time, the Prime Minister stated that the central government would provide full support for rescue and reconstruction. Emergency goods such as clothes and commodities, for example, large quantities of waterproof sheets, were provided through contracts with 26 merchants in Tottori (Tottori Prefecture, 2007). Necessary revision to current policies and the establishment of new ones was performed by the head of the Emergency Response Headquarter and conveyed to the central government.

To prevent population decline and the collapse of communities, Tottori Prefecture created the first law that permits funding the rebuilding of private houses using public taxes. The Law of Rebuilding House Subsidy was passed by the Tottori council and became a permanent law for future disaster recovery (Tanaka, 2004). Six years later, the National Act for Livelihood Reconstruction of Disaster Victims was revised to incorporate the contents of private housing subsidy.

For the Niigata Chuetsu Earthquake as well, an Emergency Response Headquarters was established immediately, and the Disaster Management Minister took the role of the chief of the governmental investigation team. Later, another Headquarters for the Promotion of Reconstruction and Recovery was established with the Vice Minister of the Cabinet Office as its chief. As was the case in Tottori, a Community Empowerment Support Team was established to prevent population decline and the disappearance of local communities, as well as to revitalize affected communities (Inagaki & Odagiri, 2014). In this way, collaborative recovery planning and decision making were enhanced through the direct collaboration between community and the local government.

Activities of Civilian Groups

After the West Tottori Earthquake, the Countermeasure Headquarters of the Social Welfare Organization was established to collect information about damage, the lives of solitary elderly people, and the need for volunteers. At the early stage, needs for volunteers focused on duties that did not require advanced skills, such as removing rubbish or emergency repair of buildings. As the experiences of volunteer coordination from the Great Hanshin Earthquake were utilized, a balance between the needs of recovery effort and the support provided by volunteers was achieved. Public Relations (PR) activities were also carried out to attract more volunteers from outside (Tottori Prefecture, 2007).

In Niigata, the Volunteer Center Headquarters was established owing to cooperation between the Niigata government, the prefectural Social Welfare Organization, the Japanese Red Cross Society, and NPOs. The experiences from the Great Hanshin Earthquake were utilized when allocating temporary housing by considering the original community–member relationships. To provide long-term support for recovery, the Community Empowerment Support Team was re-envisioned into numerous Community Empowerment Support Consultants in 2007 (Niigata Prefecture, 2015). The duties of consultants were to organize various events for empowering communities with local people and to coordinate support provided by governments and outside organizations.

Victims' needs were fully considered, and countermeasure policies were changed to adapt to the objectives of promoting the individual ability of the victims themselves. Furthermore, projects controlled by local governments were also opened to the Community Empowerment Support Consultant to include more stakeholders to participate in recovery processes.

With poor conditions of the temporary houses, problems of mental health such as loss of appetite and unwillingness to prepare food emerged. To cope with these problems, the prefectural government cooperated with various stakeholders such as the affected areas, universities, and mental health organizations to run hotlines for mental health consulting and send specialists to affected areas. Achievements of health support were seen, as reports reflected no significant declines of mental health (Niigata Prefecture, 2015).

Contributions from Community Members

In Tottori, more than 40% of the affected victims were elderly, illustrating that these areas were seriously depopulated. Even before the earthquake, economic and social activities in these areas were stagnant. However, after the earthquake, through collaboration with outside volunteers and local governments, the situation in some areas was significantly enhanced. For example, the percentage of neighborhoods with *jishubo* in Tottori Prefecture rose from 57% in 2001 to 77.9% in 2013 (Cabinet Office, 2015a).

In Hino Town, Tottori, the Hino Volunteer Network was established to specifically support the activities of the *jishubo* and the Disaster Volunteer Center (Hosoda, 2001). Its main objectives were to provide solutions for the problems of elderly people, to communicate with elderly people, to organize events such as birthday parties, and to collect and provide information for professional organizations. Such organizations functioned well as an intermediary for the *jishubo* and the Disaster Volunteer Center and therefore became an important fixture in communities.

In Niigata, several communities lost more than 60% of their population after the earthquake, due to population relocation (Niigata Prefecture, 2015). Therefore, the aforementioned policies such as the creation of Community Empowerment Support Consultants had the purpose of motivating residents themselves to rebuild local communities. Because of the earthquake, the chances of communicating with outside volunteers were hugely increased, which helped community members to realize the importance of their natural environment, high-quality agricultural products, and traditional cultures. These interactions encouraged community members and enhanced their confidence in rebuilding.

Other policies, such as the Committee of Chuetsu Recovery Citizenship, functioned well to connect affected areas with administrative and outside supporters. Later, the Conference for Communication of Community Recovery was developed into a well-known network for recovery (Inagaki & Odagiri, 2014). Thus, the percentage of *jishubo* rose from 19.9% in 2001 to 81% in 2013 (Cabinet Office, 2015a), indicating that the dramatically enhanced risk awareness of community members would have been greatly influenced through recovery experiences.

Widespread Disasters

Recent large-scale disasters throughout the world have affected wide regions and greatly increased the complexity of the recovery process. Affected areas that contain both urban and rural zones require many more resources, supportive members with various special skills, and attention paid to diverse community members. In Japan, such widespread disasters could be represented by the Tohoku Earthquake and Tsunami that occurred in 2011, and a predicted Nankai Trough Earthquake that is expected, with 70% probability, to occur within the next 30 years (Headquarters for Earthquake Research Promotion, 2013).

Damage

The Tohoku Earthquake and Tsunami was a magnitude 9.0 undersea earthquake off the northeast coast of Japan that occurred on March 11, 2011. The earthquake triggered a powerful tsunami that reached heights of up to 40.5 m and travelled up to 10 km inland (Japan Meteorological Agency, 2011). According to reports by the National Police Agency (2016), this disaster caused 15,889 fatalities with 2,594 people still missing, and it completely or partially destroyed 401,567 houses. The damage was extensive and varied, but the enormous human loss is clearly most significant. Over 90% of fatalities were due to the tsunami (National Police Agency, 2016), reminding people living along coastal areas of the importance of quick evacuation to higher ground before a tsunami arrives.

Prompted by the Tohoku Earthquake and Tsunami, the Cabinet Office immediately organized several institutions specifically to discuss appropriate countermeasures for a Nankai Trough Earthquake. Reports from these institutions suggest that in the worst-case scenario, 1015 km^2 of land will be inundated, an area 1.8 times larger than in the Great East Japan Earthquake. Disaster experts predict that 124 municipalities would be inundated by a 5-m tsunami, and an additional 23 municipalities would be inundated by a 10-m tsunami. The highest projected tsunami could exceed 34 m, resulting in up to 323,000 fatalities (Cabinet Office, 2012).

Organizations and Policies

Four minutes after the Tohoku Earthquake and Tsunami, the Countermeasure Office had already been organized by the Prime Minister. Later, the Emergency Response Headquarters of Tohoku Earthquake and Tsunami was established, and the first conference was organized by the Chief Cabinet Secretary, within 1 hour after the earthquake (Cabinet Office, 2015b). Thus, the processes of emergency response, reconstruction, and recovery were fully supported by the central government.

A feature of the recovery process of this earthquake was that the Basic Recovery Act for Tohoku Earthquake and Tsunami was passed through parliament on June 24, 2011 (Cabinet Office, 2015b), due to the huge human and property losses, the damages caused by the earthquake and tsunami, and an unexpected nuclear incident, which were all difficult to cope with. The objectives of the Act were to promote reconstruction and recovery processes, to motivate individuals' independence activities, and to cope with depopulation and aging problems. Basic countermeasures of the Act were to secure sufficient funding, to issue national bonds, to reconstruct special designated zones, and to establish a temporary Reconstruction Agency in the Cabinet Office.

An Investigative Committee for a Nankai Trough Earthquake was established after the Tohoku Earthquake and Tsunami on August 28, 2011 (Cabinet Office, 2012). From then on, countermeasures for a Nankai Trough Earthquake were established under the supervision of the Cabinet Office. Regions were specifically designated as disaster-prone areas and were encouraged to conduct their local DRRs. Basic plans for promoting disaster countermeasures and carrying out emergency earthquake-response activities were established. A limited number of committees and working groups were organized to specifically consider countermeasures for the earthquake.

Activities of Civilian Groups

According to the National Social Welfare Association's report (2015), the number of disaster volunteers registered through the Disaster Volunteer Center was about 1 million in 2011. However, the actual numbers of volunteers likely greatly exceeds this figure. For example, during the Golden Week holiday in May 2011, at least 3,000 volunteers registered through Ishinomaki Disaster Volunteer Centers in Miyagi; however, owing to the limited ability of the volunteer center to handle applications, many other volunteers went unregistered (Ishinomaki Social Welfare Organization, 2011).

One of the most notable features of volunteer activities was that social networks built over the Internet functioned well during the different recovery stages (Kitamura, 2014). These networks were used to conduct safety confirmation activities, allocate emergency goods, and collect information such as on transportation and roads, governmental support, and recruitment of volunteers.

Because of the serious tsunami damage, electricity and water services in some locations did not recover until 2 months after the earthquake, which hugely hindered the establishment of Disaster Volunteer Centers. Because little information was available for volunteers to access, social networks became important tools to balance volunteer support and needs. Meanwhile, the predecessor organization of the Reconstruction Agency collaborated with the private service, 3.11 Reconstruction Aid Information Portal Tasukeai Japan, to collect various nonstandardized pieces of information related to volunteer activities and edit them into a standard format to motivate people to share support information (Reconstruction Agency, 2015).

At present, large-scale volunteer activities are not involved in promoting DRR. Nonetheless, based on the lessons learned from the Tohoku Earthquake and Tsunami recovery efforts, collaboration among different volunteer organizations has been performed as a primary step across wide regions to discuss how to provide quick and efficient support that is tailored to individual communities. Widespread collaboration among different local governments, such as regarding acceptance of disaster victims, organization of shelters, and information sharing through emergency systems, is also underway.

Contributions from Community Members

Contributions from disaster victims after the Tohoku Earthquake and Tsunami were especially necessary during the recovery process; as hundreds of thousands of communities were completely destroyed by the tsunami, how to rebuild these communities became a priority. Some communities began to convene *chonaikai* (a Japanese local community organization) members to discuss issues of community relocation and community mergers. Specific agendas were discussion of community design, land use plans, arrangement of facilities, and the establishment of Assemblies of Community Development, with various stakeholders.

According to the reports from local communities (Hirano et al., 2014), the most important matter in the process of emergency response, reconstruction, and recovery was mutual aid. Some communities called for internal volunteers to make community kitchens and run temporary shelters. DRR manuals that had been discussed and issued through collaboration between community residents and administrations were reported to have functioned well. Networks of both insiders and outsiders were important to collecting rescue information, commodities, and support. Therefore, some communities actively shared food, blankets, and even living spaces. The responsibility of public safety in temporary shelters was shared between community members and local police.

Local communities that expected to be seriously hit by a giant tsunami due to a Nankai Trough Earthquake put more effort into their DRR. By collaborating with experts, community members developed and conducted new types of evacuation drills such as nighttime drills and single-person drills (an evacuation drill specifically planned for one participant) (Sun et al., 2014). With data collected from these human activities, an animation of human tsunami evacuation was made and provided by researchers to help residents better understand tsunami information (Nakai & Hatayama, 2013).

According to lessons learned from the Tohoku Earthquake and Tsunami, widespread disaster must be addressed with collaboration between different regions. Therefore, large-scale joint evacuation drills have also been planned and conducted. For example, Osaka has organized widespread drills four times, which aimed at attracting 8.8 million people to participate (Osaka Prefecture, 2015), and Kanagawa has similar events that attracted more than 1 million participants (Kanagawa Prefecture, 2015). The collaboration between communities and schools is also increasing. Events include community-based disaster map-making, community kitchen cooking drills, maintenance of community shelters, and community walks to check for dangerous places and local resources.

DISCUSSION

The aforementioned four case studies clearly demonstrate the diversity of the recovery process. To enable systematic discussion about recovery, this section focuses on three of its facets: continuity, creativity, and applicable aspects.

Continuity of Policies and Activities

Because of national laws regarding disaster preparedness and responses, related organizations are required by law to be established immediately after an emergency occurs. As in the aforementioned case studies, Emergency Response Headquarters are usually established quickly after a large disaster occurs, with the governor as head.

However, continuity of recovery policies has not been functionally realized. For example, the Law of Rebuilding House Subsidy made by Tottori Prefecture in 2001 was not incorporated into the national law until 2007. Although situations could be very diverse in different regions, the difficulty of enacting local policies into national laws hugely hinders recovery processes.

Evidence of this can be seen in interview results from flood victims in Sayo, Hyogo, Japan, in 2009 (Sun, 2010). Many flood victims stated that the administrations have responsibilities to take care of affected citizens, which should include providing subsidies to rebuild private houses lost in disasters. However, the Sayo government announced its decision to provide no such subsidies, according to the policy of the Phoenix Mutual-aid Program established after the Great Hanshin Earthquake. The program has the characteristics of an insurance company, which aims to encourage normal citizens to co-establish a fund for disaster recovery, with benefits given only to those people who have invested in the program. From these case studies, it is clear that DRR and recovery policies vary due to independent political policies. This could be one important reason why Japan does not have a national-level disaster-recovery law.

Compared with the lack of continuity of government policies, civilian group activities were very continuous, and experiences were handed down and applied to other disaster-recovery activities. The importance of coordinating volunteers was greatly realized and the establishment of volunteer centers through collaboration with social welfare organizations has been established. Disaster volunteer groups and their activities continued to be refined and grow in popularity, now involving more and more people. In the recovery process of the Tohoku Earthquake and Tsunami, a standout way that the volunteer groups progressed was that after the initial period of emergency response, the volunteer groups became NPOs to continue providing support for development in affected areas.

As mentioned earlier, experiences in volunteer activities were specifically noted by various stakeholders. In the organization of shelters, volunteers provided help in distributing goods and commodities and in cleaning for disaster victims. However, compared with widespread disasters, disasters that occur in a limited area might have difficulty in attracting volunteer attention, despite serious damage.

For example, a recent giant flood that occurred in Ibaraki, Japan, destroyed a large amount of houses and caused other such property loss (Commons, Ibaraki NPO Center, 2015). However, there were very few volunteers, as the Social Welfare Organization and local governments lost basic function due to serious damages, and because the affected area was limited to one small city in Ibaraki, meaning that the city did not greatly attract volunteers. Increasing awareness of disaster damage and the motivations of disaster volunteers should be considered important possible areas of future research.

Creative Countermeasures

Through the summary of recovery policies in Japan, it is clear that the existing policies and rules do not appropriately address the circumstances of later disasters. In the case study of the West Tottori Earthquake, the establishment of the Law of Rebuilding House Subsidy was specifically created for the affected areas (Tanaka, 2004). The basic consideration about public money allocated to private housing was that it costs about 3 million JPY to construct a temporary house that has to be destroyed in 2 years. Considering the aging demographics of the local population, this would result in rapid local population decline. Conversely, by providing 1 million JPY to support reconstruction of a private house, residents would be encouraged to come back to rebuild.

Survey data (Takahashi, Nakamura, & Matsumoto, 2002) indicated that the affected victims highly valued the law, with comments such as the high amount of subsidy allocated to house repairs. As there was no age limitation for subsidy applications, higher value came from young adults. More than 30% respondents answered that without the building subsidy, they would probably not have been able to rebuild their houses.

After the Niigata Chuetsu Earthquake, Community Empowerment Support Consultants were established specifically for coping with seriously depopulated earthquake-affected areas. In the recovery process of the Tohoku Earthquake and Tsunami, 452 consultants are working actively from 21 organizations in the affected prefectures of Miyagi, Fukushima, and Iwate (Reconstruction Agency, 2015). The establishment of the Reconstruction Agency was a historically novel action in the recovery process. The agency controls issues in many fields, such as the recovery situation, recovery activities, support for affected victims, and volunteer activities.

Applicable Aspects of Disaster Recovery

According to survey data from the National Social Welfare Association (2015), support from outsiders tends to sink below the level of insiders at 1 year after a disaster. Disaster-affected people who had frequently participated in *chonaikai* activities showed a higher chance of receiving supports from outsiders and of providing support to other community members.

Therefore, in the DRR process for a Nankai Trough Earthquake, establishment of the Community-based Disaster Management Plan in 2015 was largely taken from lessons learned from the recovery process of the Tohoku Earthquake and Tsunami. The plan specifically focuses on the independence of local communities, aiming to motivate every resident to participate in community development and DRR. In Kuroshio Town, Kochi, the local government has dispatched all governmental staff to communities for providing assistance in making disaster-readiness plans (Kuroshio Town, 2015).

Whether DRR or disaster recovery, community-based activities encourage people to become independent, confident, and eager to make decisions and take responsibility for themselves. The roles of government, outside volunteers, and other stakeholders are to provide necessary and expected support but not to replace or suppress the independent ability of community members.

CONCLUSION

This chapter discussed several case studies of disaster recovery in Japan. Many past disasters have provided valuable evidence and uncovered great challenges for future policies regarding disaster recovery. To reduce property damage and human loss in future natural disasters, such as a Nankai Trough Earthquake, it is important for Japan to take more actions on DRR based on its past experiences.

As has been discussed, even in domestic Japan, recovery policies could differ greatly from prefecture to prefecture. There are weaknesses and strengths to the current recovery system, but the situation is being gradually changed to provide better support to disaster victims. Owing to Japan's frequent natural disasters, the country has accumulated experience about the recovery process; however, the consequences of natural disasters in Japan have also become more and more complex. Thus, policies related with DRR and recovery must be revised and verified further to empower affected communities. We hope that Japan can provide its experiences with disaster recovery to other countries, especially those in Asia that face similar natural disasters and that these countries can apply our findings in a way that is appropriate to their local communities through in-depth investigation into societal, cultural, and systematic issues.

REFERENCES

Architectural Institute of Japan. (Ed.). (1995). *Hyogo ken nanbu jishin saigai chosa sokuho* [Report on the Great Hanshin Earthquake]. Tokyo: Architectural Institute of Japan.

Cabinet Office. (2012, August 29). *Nankai trough kyodai jishin no higai sotei ni tsuite* [Damage estimates of a Nankai Trough Earthquake]. Retrieved from http://www.bousai.go.jp/jishin/nankai/taisaku/pdf/20120829_higai.pdf

Cabinet Office. (2015a, June). *Bosai hakusho* [Disaster white paper]. Retrieved from http://www.bousai.go.jp/kaigirep/hakusho/index.html

Cabinet Office. (2015b, September 9). *Higashi nihon daishinsai kanren jyoho* [Information related with the Tohoku Earthquake and Tsunami]. Retrieved from http://www.bousai.go.jp/2011daishinsai/

Chamlee-Wright, E. (2011). Social capital as collective narratives and post-disaster community recovery. *The Sociological Review*, *59*(2), 266–282.
Commons, Ibaraki NPO Center. (2015). *Special edition of Joso flood*. Ibaraki: Commons, Ibaraki NPO Center.
Cretney, M. R. (2016). Local responses to disaster: The value of community led post disaster response action in a resilience framework. *Disaster Prevention and Management*, *25*(1), 27–40.
Headquarters for Earthquake Research Promotion. (2013, May 24). *Nankai trough no jishin katsudo no choki hyoka (daini ban) ni tsuite* [Estimation of seismic activities of Nankai Trough Earthquakes]. Retrieved from http://www.jishin.go.jp/main/chousa/13may_nankai/index.htm
Hirano, T., Ogiso, S., Kodama, Y., Hosaka, M., & Ikeda, M. (2014). Issues and prospects in post-disaster support activities for affected families of the Tohoku Earthquake: A community strengthening perspective toward self-reliant livelihood. *Journal of Social Welfare, Nihon Fukushi University*, *130*, 67–88.
Hosoda, K. (2001). *Hino-cho ni okeru volunteer activity ni tsuite* [Volunteer activities in Hino Town]. *Journal of Natural Disaster Science*, *20*(3), 242–244.
Hyogo Prefecture. (2016, January 26). *Shinsai fukko* [Catastrophe recovery]. Retrieved from https://web.pref.hyogo.lg.jp/town/cate2_206.html
Inagaki, F., & Odagiri, T. (2014). *Shinsai fukko ga kataru nosanson saisei: Chiiki-zukuri no honshitsu* [Revitalization of farming and mountain villages from disaster recovery: The essential of community development]. Tokyo: Commons.
Institute of Social Science, Waseda University. (Ed.). (1996). *Hanshin awaji daishinsai ni okeru saigai volunteer katsudo* [Activities of disaster volunteer in the Great Hanshin Earthquake]. Tokyo: Institute of Social Science, Waseda University.
Ishinomaki Social Welfare Organization. (2011, May). *Volunteer ukeire jyokyo* [Registered volunteers]. Retrieved from http://svc.miyagi.jp/a311/index.php?module=blog&blk_id=16320&page=2
Japan Meteorological Agency. (2011, April). *2011 nen tohoku chiho taiheiyo oki jishin ni yoru kakuchi no shindo* [Seismic intensity in areas affected by the Tohoku Earthquake and Tsunami]. Retrieved from http://www.data.jma.go.jp/svd/eqev/data/2011_03_11_tohoku/0311_shindo.pdf
Japan Meteorological Agency. (2016). *Earthquake database*. Retrieved from http://www.jma.go.jp/jma/menu/menureport.html
Kanagawa Prefecture. (2015, October 1). *Kanagawa shakeout*. Retrieved from http://www.pref.kanagawa.jp/cnt/f480078/
Kawata, Y. (1997). Prediction of loss of human lives due to catastrophic earthquake disaster. *Journal of Natural Disaster Science*, *16*(1), 3–13.
Kitamura, T. (2014). Higashi nihon daishinsai ni okeru volunteer katsudo to jyoho [Volunteer activities and information in the Tohoku Earthquake and Tsunami]. *Disaster Preparedness for Fire Safety and Information*, *115*, 29–32.
Kobe City. (2014, December 10). *Hanshin awaji daishinsai no kiroku* [Record of Great Hanshin Earthquake]. Retrieved from http://www.city.kobe.lg.jp/safety/hanshinawaji/index.html
Kuroshio Town. (2015, December 24). *Bosai kankei, Kuroshio-cho no keikaku* [Disaster risk reduction plan of Kuroshio Town]. Retrieved from http://www.town.kuroshio.lg.jp/pb/cont/bousa-kankei/499
Nakai, F., & Hatayama, M. (2013). Analysis of residents' evacuation behavior and investigation of evacuation plans by residential cooperation: Case study of disaster risk communication

at Kuroshio Town, Kochi Prefecture. *Proceedings of Civil Engineering Planning, 47*, CD-ROM (54).

National Police Agency. (2016, January 8). *Higashi nihon daishinsai ni tsuite* [The Tohoku Earthquake and Tsunami]. Retrieved from http://www.npa.go.jp/archive/keibi/biki/index.htm

National Social Welfare Association. (2015, December 31). *Saigai volunteer center de uketsuketa volunteer katsudo shasu no suii* [Registered volunteers at Disaster Volunteer Center]. Retrieved from http://www.saigaivc.com/

Niigata Prefecture. (2015, March 25). *Niigata ken tyuetsu daishinsai fukko kensho hokokusho* [Verification report of Niigata Tyuetsu Earthquake]. Retrieved from http://www.pref.niigata.lg.jp/shinsaifukkoushien/1356808376486.html

Osaka Prefecture. (2015, December 24). *Osaka 880 mannin kunren* [Evacuation drill of 8.8 million participants in Osaka]. Retrieved from http://www.pref.osaka.lg.jp/shobobosai/trainig_top/

Paton, D. (2007). Preparing for natural hazards: The role of community trust. *Disaster Prevention and Management, 16* (3): 370–379.

Reconstruction Agency. (2015). *Volunteer, NPO, and public-service corporation*. Retrieved from http://www.reconstruction.go.jp/topics/main-cat4/sub-cat4-2/

Springgate, B., Wennerstrom, A., & Carriere, C. (2011). Capacity building for post-disaster mental health since Katrina: The role of community health workers. *Review of Black Political Economy, 38*, 363–368. doi:10.1007/s12114-010-9083-x

Statistics Japan. (2016). *Japan statistical yearbook*. Retrieved from http://www.stat.go.jp/data/nenkan/index1.htm

Sun, Y. (2010). *Building social resilience through disaster risk reduction* (Unpublished master's thesis). Beijing Foreign Studies University, Beijing.

Sun, Y., Yamori, K., & Kondo, S. (2014). Single-person drill for tsunami evacuation and disaster education. *Journal of Integrated Disaster Risk Management, 4*(1), 30–47. doi:0.5595/idrim.2014.0080

Takahashi, K., Nakamura, S., & Matsumoto, R. (2002). Relief system about reconstruction of dwellings in the 2000 Western Tottori Earthquake and resident's receipt. *Reports of Graduate School of Engineering, Nagasaki University, 32*(59), 139–145.

Tanaka, N. (2004). *Kaikaku no gijyutsu: Tottori ken chiji Katayama Yoshihiro no chosen* [Revolution skills: Challenges from the Tottori governor of Katayama Yoshihiro]. Tokyo: Iwanami Shoten.

Tottori Prefecture. (2007). *Tottori ken seibu jishin kankei* [Archives of West Tottori Earthquake]. Retrieved from http://www.pref.tottori.lg.jp/seibujisin/

Umeda, Y. (2001). *Tottori ken seibu jishin ni yoru saigai ni kansuru chosa kenkyu* [Report of West Tottori Earthquake]. Research Report Funded by 2000 Grants-in-Aid. Kyoto University, Kyoto.

United States Geological Survey. (2016). *Earthquake facts and statistics*. Retrieved from http://earthquake.usgs.gov/earthquakes/eqarchives/year/eqstats.php

Weissbecker, I. (2009). Mental health as a human right in the context of recovery after disaster and conflict. *Counselling Psychology Quarterly, 22*(1), 77–84.

CHAPTER 6

CITIZEN PARTICIPATION IN THE DISASTER RECONSTRUCTION PROCESS: LESSONS FROM THE GREAT EAST JAPAN EARTHQUAKE

Takashi Tsuji

ABSTRACT

Citizen participation has attracted attention in the context of decentralization. In a disaster reconstruction process, a business plan for reconstruction can be modified in line with diversified situations of disaster-affected areas by citizen participation. In Japan, the central government makes a decision about the authority in charge of an overall disaster reconstruction and the budget planning, whereas local governments are in charge of creating and implementing a business plan for reconstruction of each local municipality. Therefore, local governments play an important role in organizing citizen participation to realize the reconstruction that fits reality. It has yet to be shown as decentralization reform and citizen participation system in Japan produce the socio-spatial inequality after the Great East Japan Earthquake. However, it remains to be elucidated how local

Recovering from Catastrophic Disaster in Asia
Community, Environment and Disaster Risk Management, Volume 18, 105–126

ISSN: 2040-7262/doi:10.1108/S2040-726220160000018005

government and community have to operate the institution about citizen participation during the disaster reconstruction process. I have been doing fieldwork on three tsunami-affected sites in Miyagi Prefecture over past 4 years: Onagawa Town, Higashimatsushima City, and Natori City. I have investigated the social processes of making and implementing a reconstruction plan, and citizen participation. The findings from my fieldwork are as follows: First, citizen participation is based on organizing residents at the community level. Second, traditional community organization (such as neighborhood organization abd industrial associations) contribute to organize residents especially in the emergency phase. Third, as the disaster phase moves, local government and community organization need to change the previous participation frame to ensure residents representation and policy legitimacy.

Keywords: The Great East Japan Earthquake; citizen participation; community organization

INTRODUCTION

Disasters are political phenomena (Olson, 2000). During the distribution of relief goods and disaster reconstruction projects in regional communities struck by a disaster, governments coordinate the interests of the various actors. However, because government reconciliation of interests in disaster situations is conducted with limited time and unreliable information (Lidskog & Sjödin, 2015), often there is antagonism among interested parties over procedures and policies. Disasters expose issues of regional community governance and become opportunities to question the validity and legitimacy of the governing system (Ahrens & Rudolph, 2006; Davis, 2005; Tierney, 2012).

One point of contention about regional community governance in recent years is reputedly local-level citizen participation (Lombard, 2013). In Japan, this corresponds with the decentralization reforms that have been taking place since the 1990s. The Japanese government has promoted two methods of such reform: municipal mergers and the devolution of power to local communities.

However, local-level citizen participation in Japan was not uniformly established by the government through legislation[1] but was instead determined and promoted by local governments. Therefore, there are a variety of the institutionalization of citizen participation depending on the condition of local communities (Satoh, 2007). For example, in the 1990s, while some local governments established new community organizations and regulated the organizations' financial and political authority, other local governments did not institutionally reform community organizations. The Great East Japan Earthquake of March 11, 2011, occurred just as decentralization reforms and local-level citizen participation were making progress.

In the reconstruction after the Great East Japan Earthquake, in addition to taking charge of a majority of the reconstruction budget, the central Japanese government created institutions for managing and using reconstruction budgets (reconstruction grants)[2] and then positioned 40 projects contributing reconstruction grants as "substantive projects necessary for regional development."[3] Because of this, Japan's disaster reconstruction system had large limitations to reconstruction budgets and projects built into it by the government. Nonetheless, although there was a precedent of limitations on government-backed budgets and projects, it was the local governments and communities that had the final say as to what kinds of reconstruction plans to formulate and whether or not to expand the reconstruction projects.

In the aftermath of the Great East Japan Earthquake, local governments incorporated citizen participation and formulated reconstruction plans and continue to run reconstruction projects. Through citizen participation, local governments can formulate plans that suit the conditions of the disaster-struck areas and operate reconstruction projects (Edgington, 2010). Citizen participation thereby becomes a governance tool in the disaster reconstruction process. However, since pre-earthquake government-run citizen participation institutions differed geographically, there were regional differences in the levels of citizen participation in the reconstruction process.[4]

Cho points out that in the Great East Japan Earthquake reconstruction there is a top-down policy formation by government leadership, and citizen participation was lacking. Furthermore, the changes to citizen participation that accompanied the pre-disaster devolution of power to local communities allowed for the possibility of socio-spatial inequalities in reconstruction (Cho, 2014). However, regarding citizen participation, Cho mainly focuses on the drafting of the budgets of national, prefectural, and local governments and pays little attention to citizen participation in local government policy making or the procedures of citizen participation in local communities.

We conducted a comparative analysis of three local governments in Miyagi Prefecture, all of which suffered enormous tsunami damage in the Great East Japan Earthquake and had different experiences and paces of reconstruction. First, we analyzed the formative history of community organizations in these pre-disaster regional communities. Then, we examined citizen participation policies that the local governments adopted when formulating reconstruction plans and running reconstruction projects, and the role of community organizations in the process.

RESEARCH DESIGN

Institutional Approaches to Disaster Reconstruction Research

In recent years, researchers have been focusing on institutions to clarify the primary determining factors in the success or failure of disaster reconstruction in regional communities. Researchers have established that local governments and communities have a tendency to promote reconstruction on the basis of pre-disaster policies and planning (Güiza, Simmons, Burgess, & McCall, 2016). When analyzing social disaster reconstruction processes, researchers are starting to recognize the effectiveness of an institutional approach that focuses on local institutional arrangements before and after the disaster (Smith & Birkland, 2012). Scholars have focused predominately on two dimensions regarding institutional approaches.

The first is a temporal perspective on pre-disaster inherent institutions and post-disaster adoptive institutions. While local governments and communities inherit pre-disaster institutions after disaster strikes, they also create new institutions in post-disaster social processes (Hilhorst, 2013; Tierney, 2014). Pre- and post-disaster institutions can coexist, and there is the possibility of both functioning with overlapping duties.

In the context of Japan's disaster reconstruction, post-earthquake town reconstruction councils are cited as post-disaster adoptive institutions. When the Great Hanshin-Awaji Earthquake happened in 1995, post-earthquake town reconstruction councils were formed in every district, and about 100 organizations took action (Nozaki, 2005). These councils were established with the goal of reflecting residents' demands for land readjustment projects. In areas struck by the Great East Japan Earthquake as well, post-earthquake town reconstruction councils act as a space for consensus building for local governments and communities concerning the operation of reconstruction projects. Some councils were established on the basis of pre-disaster community organizations, while others were established independent of existing organizations (Iwasaki, 1998). In areas struck by the Great East Japan Earthquake, the coexistence of new and old community organizations can be seen in the disaster reconstruction process. In this chapter, I will focus on the reorganization process of pre- and post-disaster community organizations.

The second is a public–private perspective on public institutions belonging to local governments and local community organizations. In the context of this research, the former refers to all pre-disaster local government policies, and post-disaster reconstruction and citizen participation policies. In the latter, researchers identify the importance of the role played by community organizations in reconstruction. For example, Daly positions community organizations as the political infrastructure of regional communities and points out that community organizations performed intermediary functions between national and local governments with regard to the distribution of relief supplies and the formulation of reconstruction plans (Daly, 2014). Many researchers highlight the role of traditional community organizations (Daly, 2014; Husin & Alvisyahrin, 2014).[5] In this chapter, I will situate traditional community organizations, active locally before the disaster, as institutional foundations for citizen participation, reflecting the policy making process of local governments.

Research Questions and Observation Points

The main objective of this paper is to seek the contributing factors that produce regional differences in the process of disaster reconstruction. The dependent variable, regional diversity, is assumed to be the pace of disaster reconstruction. The explanatory variables and hypotheses will be the local government and the community institutions that form the bases of citizen participation in the regional communities. To examine the validity of my hypotheses, We established the following two observation points with reference to classical community types (Berke, Kartes, & Wenger, 1993).

The first point targets is the relationship of vertical reconciliation of interests between local governments and community organizations. In Japan, the levels of institutionalization of citizen participation differed by region even before the disaster. As such, it is thought that local government preferences for intervention in community organizations and the function of advocacy by community organizations are varied.

The second is the horizontal relationship between community organizations. In areas affected by tsunami damage from the Great East Japan Earthquake, several community organizations coexisted before the disaster. In the reconstruction period, the arrangement of community leaders and citizen representation in communities has continued to be influenced by decentralization reforms and has differed by region.

On the basis of the points, we considered the formulation process of community organizations in pre-disaster local governments and the responses of community organizations to citizen participation in the disaster reconstruction process.

RESEARCH METHODS AND FIELDWORK AREAS

Research Methods

In this research, we applied qualitative research methods. We have been conducting fieldwork in this area since February 2012. Fieldwork involved conducting interviews with the reconstruction management department staff of local governments and the leaders of community organizations (industry groups, local resident organizations, etc.), gathering outlines of citizen participation policies in local government reconstruction processes, and collecting data about the participation of community organizations in reconstruction policies. In addition, we observed meetings and conferences held by local governments and community organizations concerning reconstruction. In assembling the formative history of local community organizations, we collected pre-disaster archival materials, such as local governmental public relations documents and newspaper articles.

Areas of Fieldwork

The areas of fieldwork covered in this research are three locations in Miyagi Prefecture in the Tohoku region: Onagawa Town, Higashimatsushima City, and Natori City. Miyagi Prefecture contains Sendai, the largest city in the Tohoku region. As the front line of post-World War II Tohoku region development, it drove the politics and finances of the Tohoku region long before the Great East Japan Earthquake disaster. Development in the Tohoku region went hand-in-hand with the development of large-scale urban infrastructure. The places of fieldwork for this study include the Sendai Bay New Industrial Cities (1962), which were born out of a 1960s development project and are areas that aimed for coexistence with large-scale urban infrastructure before the disaster.[6]

Regarding the decentralization reforms in the 1990s, of the three fieldwork areas, only Higashimatsushima carried out a municipal merger before the Great East Japan Earthquake. Higashimatsushima City was formed in 2005 from the merger of Yamoto Town and Naruse Town. However, in the 1990s, the local governments and residents of Onagawa Town and Natori Town too considered municipal mergers. Onagawa Town looked into a merger with the adjacent Oshika Town in 2002 but withdrew its proposal in 2003, foreseeing a "lack of economic benefit from the merger" (*Kahoku Shimpo*, 2003). Onagawa Town had been in receipt of subsidies and property taxes from the Onagawa nuclear power plant and thus lacked the financial incentive for a municipal merger that ordinary local governments have. Natori City considered merging with Sendai City in the 1990s, and its citizens split into supporting and opposing factions and began petitioning campaigns. However, in 1995, the Natori city council rejected a citizen proposal seeking the establishment of a merger congress with Sendai City, and withdrew its merger proposal (*Asahi Shimbun*, 1995).

Since the earthquake disaster, Miyagi Prefecture has promoted "creative reconstruction." It consists of industrial structural reform (such as the consolidation and expansion of agricultural land, and the establishment of special reconstruction district for the fisheries industry), and the development of residential infrastructure concomitant with location. The three places studied in this chapter are each formulating reconstruction plans following Miyagi Prefecture's reconstruction objectives and operating reconstruction projects.

The locations of the fieldwork areas for this study are shown in Fig. 1, and their respective populations, industries, tsunami damage, reconstruction plans, and project formulation phases are detailed in Table 1.

HISTORY OF THE FORMATION OF LOCAL ORGANIZATIONS

Central District, Onagawa Town

Community organization in Onagawa Town was closely related with the Onagawa nuclear power plant. The Onagawa town council voted over whether to invite the construction of a nuclear power plant in the town in 1967. At the time, the local government did not hold discussions with the fishery cooperatives, which were the interested parties regarding fishing rights reparations, about the power plant (Watanabe, 1999). The local government made an error in the political procedure; and, as a result, the anti-nuclear movement, with the fishery cooperatives at its core, intensified into the 1980s.

Onagawa Town of the 1960s enjoyed an abundance of deep sea and offshore skipjack tuna fishing, and fishery cooperative union membership was high. Nuclear power became a point of contention in the 1967 mayoral

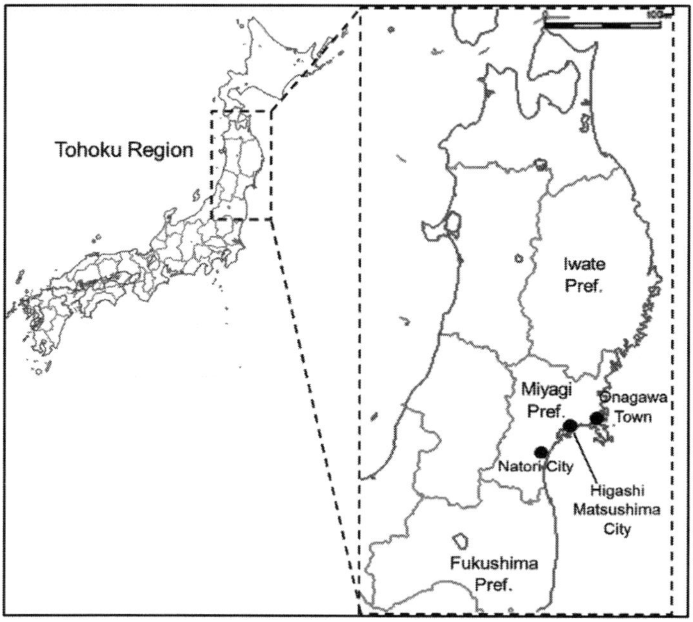

Fig. 1. Map of Fieldwork Areas. *Source*: Made by the author with reference to the Teikoku-Shoin outline map, Cho (2014, p. 159).

election, and the Onagawa Fishery Association nominated its union president as a candidate due to its opposition to the nuclear power plant. Although the union president ultimately lost the mayoral election, this opposition candidacy won 4195 votes, closely contesting the (nuclear supporting) incumbent's 5190 votes (*Ishinomaki Kahoku*, 1983). At that time, the fisheries cooperative had a large influence on town administration.

After the construction of the Onagawa Nuclear Plant was decided in the 1980s, Fishery Association union membership dwindled with the impact of a slump in deep sea skipjack tuna fishing.[7] In place of the Fishery Association, the community organizations that arose to promote their collective interests to the local government were "purchaser cooperatives" and a "chamber of commerce and industry" organized by self-employed people within the town.

From the 1970s, the pressing issue for local members of the fish-processing industry was securing source fish other than skipjack for processing. Fish-processing industry members founded the Purchasers Association in 1978. Their objective was to attract Pacific saury fishing boats from outside the

Table 1. Outline of Fieldwork Areas.

		Onagawa Town	Higashi Matsushima City	Natori City
	Population	**10,051**	**42,908**	**73,140**
Population and Industry	Employment in primary sector (%)	**15.1**	**9.1**	**4.4**
	Employment in secondary sector (%)	**32.3**	**25.4**	**22.2**
	Employment in services (%)	**52.0**	**65.4**	**73.1**
Tsunami Damage	Death Toll	**872**	**1,152**	**993**
	Death Rate (%)	**8.68**	**2.68**	**1.36**
	Property Damage	**3,273**	**11,077**	**3,930**
Reconstruction Plan and Project	Development Phase of the Reconstruction Plan	**September, 2011**	**December, 2011**	**October, 2011**
	Start Phase of the Physical Infrastructure Project	**September, 2012**	**October, 2012**	**October, 2014**
	Site Area of the Physical Infrastructure Project (ha)	**226.4 (Central district)**	**91.2 (Nobiru district)**	**121.8 (Yuriage district)**

We selected and analyzed the districts whose reconstruction projects were the largest in scale (the Central district of Onagawa, the Nobiru district of Higashimatsushima, and the Yuriage district of Natori).

Source: National census (2010), Miyagi Prefecture documents.

Note: Population and industry are current as of 2010; death toll current as of December 31, 2015. Property damage is the number of structures more than partially damaged.

prefecture in order to provide source fish with a high degree of freshness to Onagawa port for processing. Upon establishment, the Purchasers Association requested a ¥120 million subsidy from the town and started to work on ice-making operations to preserve the freshness of the saury (Onagawa Fish Market Purchasers Association, 1998). After that project ended, the local government used funds from the nuclear power plant budget to continue to attract fishing boats from outside the prefecture and actively supported the Purchasers Association (*Onagawa Town News*, 1992). From the 1980s, the Purchasers Association included over 20 companies, each with annual sales of over ¥1 billion; it even expanded into social responsibility efforts such as financing a town event (the saury harvest festival). In this way, the Purchasers Association grew to be the most influential industrial organization in the town.

The Chamber of Commerce and Industry, which was interested primarily in economic results, promoted the nuclear power plant. After the plant was constructed, the Chamber of Commerce alternately cooperated with and opposed the Purchasers Associated on the use of nuclear plant subsidies and industry stimulation measures.[8] However, in 1993, the town used nuclear subsidies to promote business and sightseeing, constructing a fisheries sightseeing center.[9] Hence, the Chamber of Commerce and Industry routinely coordinated with the local government on these operations.

Nobiru District, Higashimatsushima City

In Higashimatsushima City, before the municipal merger, there were *Gyousei-Ku*, traditional local organizations in every village, which existed as community organizations overseeing all member households. In the Nobiru district, there were 12 *Gyousei-Ku* that carried out activities such as cleaning and summer festivals. Residents saw the president of their *Gyousei-Ku* as the leader of their village, and the *Gyousei-Ku* activities were supported by the president.

In Higashimatsushima City, the local government institutions regulated the roles and activities of the *Gyousei-Ku*. Their presidents had a part-time special position in government service entrusted by the local government. The local government's regulations stipulate that *Gyousei-Ku* presidents have the duty to attend *Gyousei-Ku* president meetings.[10] *Gyousei-Ku* president meetings were a forum for the local government to provide information to the *Gyousei-Ku* about its policies and management and to share information between both parties. However, excepting the issue of noise from the Japan Air Self-Defense Force Matsushima base, opportunities for the *Gyousei-Ku* to make policy proposals to the local government were limited. Therefore, in 2002, in the Nobiru district, Naruse Town introduced a "community development district leadership member institution." The goal of this institution was to reflect residents' demands to the town government by putting responsible staff members into the *Gyousei-Ku* (*Naruse Town News*, 2002). The existence of the institution implies that the local government was involved in the reconciliation of interests and intended to intervene in community organizations.

The local government took the opportunity of the municipal merger of 2005 to reorganize community organizations, of which *Gyousei-Ku* played a central part. After the merger, *Gyousei-Ku* presidents received guidance from the local government and established an all-household registration type of local organization called "Nobiru District community development council" (hereafter "Community Development Council"). As with the *Gyousei-Ku*, local public institutions regulated the roles and activities of the community development councils. Higashimatsushima City regulated the Community Development Council's subsidy contributions and authority with its Community Development Basic Regulations (December 25, 2005). Further, the city specified that the council's remit was to endeavor "to reflect the will of area self-governing organizations in policy formulation" and guaranteed opportunities for Community Development Council advocacy. In Nobiru district, all of the *Gyousei-Ku* presidents assumed positions as officials in the Community Development Council. The activities of the *Gyousei-Ku* and their presidential leadership continued even after the founding of the Community Development Council.

Yuriage District, Natori City

The community organization formation process in the Yuriage district of Natori City before the disaster was influenced by the expansion of the Sendai City metropolitan area following World War II. The Yuriage district, located in the northwest part of Natori City on the coast of Sendai Bay, had prospered as the Yuriage port area, with workplaces and homes set close together, before the war.

Community organizations in the Yuriage district featured the coexistence of diverse organizations in the same space. Organizations active in the Yuriage district, before the disaster, can be classified into two main types: industry associations and neighborhood associations.

In the first category were the Fishery Association and the Chamber of Commerce and Industry. In 1952, when the inshore fishing industry was prospering in the Yuriage district, of the 1281 households in the district, 983 were engaged in the fishing industry (Okazaki, 1977). The Fishery Association had 879 union members (1949) and was active as a community organization representative of the Yuriage district (Natori City, 1977). However, from the 1960s, young people stopped pursuing fishing after the industrialization of the Sendai City metropolitan area, and fishing operations had reduced (in number) to just 31 households by 2008. Thus, the Fishery Association contracted (Natori City, 2011). The Natori City Chamber of Commerce and Industry was founded in 1960 and established branches in eight districts in the city aside from Yuriage. By 2001, the Yuriage district Chamber of Commerce and Industry membership included 184 establishments (70 in commerce, 70 in manufacturing, and 44 in the service sector) (*Natori Commerce and Industry News*, 2001). In the 1990s, when the municipal merger became a point of contention, the Chamber of Commerce and Industry strengthened its participation in the municipal government. It sympathized with the mayor, who had displayed a cautious approach to the municipal merger, and held dialogues, thereby becoming a forum for citizen participation (*Asahi Shimbun*, 1996).

The second category is neighborhood associations, *Chounaikai*. In the Yuriage district, there were 11 *Chounaikais*, but the times of their establishment, and their staff composition and activities are not uniform. There were no *Gyousei-Ku* community organizations here. The *Gyousei-Ku* of Natori City were under the domain of *Gyousei-Ku* presidents who were responsible for supporting administrative affairs under the direction of the local government.

In the Yuriage District before the disaster, there were no community organizations whose roles or activities were regulated by local government institutions. The local government did not regulate the *Chounaikai* through their institutions. Local residents only regarded the *Chounaikai* head as a representative informally. Similarly, *Gyousei-Ku* presidents, like neighborhood association heads, were simply individual leaders among many others existing in the area.

The local government in each area, depending on the policy matter, would include industry associations as committee members and establish committees or symposia. However, these were all temporary and were not meetings that representatives from all community organizations in Yuriage district could attend.[11] Community organizations that routinely negotiated with the local government were limited to a subset of industrial associations such as the Chamber of Commerce and Industry. In sum, the interest-reconciling relationship between the local government and community organizations in Natori was not a close one.

THE DISASTER RECONSTRUCTION PROCESS AND CITIZEN PARTICIPATION

Since the disaster of March 11, 2011, Onagawa Town, Higashimatsushima City, and Natori City have promoted large-scale reconstruction projects. As of June 2016, residential infrastructure developments are complete in parts of some districts in Onagawa Town and Higashimatsushima City, and some residents are rebuilding homes in newly developed areas. The media has named Onagawa Town and Higashimatsushima City as "reconstruction front-runners," and both have been acknowledged for their quickly progressing reconstructions (Katoh, 2015; Sakurai, 2015). On the other hand, Natori City is about 2 years behind Onagawa Town and Higashimatsushima City in starting its reconstruction work (Table 1). Underlying this delay was a difficult consensus-building process between local governments, communities, and residents over reconstruction projects in the Yuriage district of Natori City.

Central District, Onagawa Town

Reconstruction planning for the Central district of Onagawa Town featured disaster prevention measures (foundation-raising) of the urbanized area within a zone of about 220 ha, with hill reclamation advancing simultaneously. The land readjustment projects sought to establish a foundation for homes and the fisheries industry, without enclosing the harbor with a seawall. The key features of this plan were not formulated by the local government but by the Onagawa Reconstruction Coordination Conference (*Fukkou Renraku Kyougikai*, hereafter referred to as "FRK"), which consisted of industry groups such as the Purchasers Association and the Chamber of Commerce and Industry. Industry groups that had a latently antagonistic relationship before the disaster seized the opportunity to cooperate on reconstruction.

The FRK was founded about 1 month after the disaster, on April 19, 2011; and the chair of the Chamber of Commerce and Industry assumed the post of chair of the FRK. The FRK established committees for every branch of industry (fisheries and commerce), summarized the results of the discussions of each committee, and made reconstruction plan proposals to the mayor and town council.

The mayor and town council reflected most of the FRK's proposals in the plans made by the local government. The FRK appeared in large numbers as officials in the town's reconstruction plan formulation committee, and the committee reviewed the FRK's reconstruction plan proposals. The FRK was the foundation for citizen participation in Onagawa in the earliest stages of recovery, fiscal year 2011, and the reconstruction plans correspondingly reflected the intentions of those involved in fish processing and local commerce.

The local government formulated its full reconstruction plan in September 2011. Average citizens became widely involved in the reconstruction process from the latter half of the 2011 fiscal year, after the plan was formulated, at which point the local government is said to have used two methods to manage citizen participation.

First, it held post-earthquake town reconstruction information sessions. As of November 2015, the town had held its 13th information session. The first round of information sessions totaled 40 meetings convened in several venues within and outside the town, with 1,380 people in attendance. At the information sessions, the town provided residents with detailed information about housing reconstruction, such as target districts and schedules for housing infrastructure renovation projects.[12]

The second method involved workshops. The local government set up the Onagawa Town Reconstruction Promotion Council in November 2011 and appointed 17 members, including officials from the FRK. From fiscal year 2012 to 2013, the Town Reconstruction Working Group, made up of people and public appeals recommended by council members, considered stationary scenery and public facilities in central districts. Since fiscal year 2014, the council has solicited participation by residents other than those recommended by council members and is diligently training new leaders in the area.

Nobiru District, Higashimatsushima City

A notable feature of the recovery plans in the Nobiru district of Higashimatsushima City was the level of interaction around the high ground relocation projects that developed residential infrastructure by reclaiming hills in the northern part of the district. In addition to the local government formulating a fundamental plan of action, the *Gyousei-Ku* presidents played a role. The Nobiru district *Gyousei-Ku* presidents appealed to the city for "general high-ground relocation for the entire district," and jointly submitted a petition to the city on May 11, 2011. The petition submitted by the *Gyousei-Ku* presidents, who were town reconstruction council members, became the basis for the local government to justify the high-ground relocation project.

However, the Nobiru District includes residents whose homes were not completely washed out by the tsunami (*zaitaku* victims). They protested against the *Gyousei-Ku* presidents' settlement, gathering 639 signatures from 222 households by June 27, 2011, and submitted a petition to the city asking to rebuild their homes in the current location. The local government accepted their petition and granted permission for citizen residences in tsunami inundation districts on the condition that they take certain disaster prevention measures, demonstrating that they could avoid making one-sided residential regulations for the tsunami inundation districts.[13] From fiscal year 2011, *zaitaku* victims organized the Nobiru District Zaitaku Resident Association (hereafter referred to as the Zaitaku Resident Association),[14] opening a path for negotiation with the local government, without mediation by the *Gyousei-Ku* or town reconstruction council, about the needs of residents.[15]

After fiscal year 2012 as well, community organization became the foundation for citizen participation in the Nobiru district. Although, depending on the progress of reconstruction projects, local governments and communities were repeatedly restructured.

In July 2012, when the town reconstruction council set up a section meeting to discuss the district's reconstruction (the reconstruction section meeting), the Zaitaku Residents Association participated. They opposed the town reconstruction council on the details of the reconstruction plan formulation; yet they participated in planning with the town reconstruction council since it is an organization that represents residents, and the local government has an obligation to reflect the demands of the council.

In fiscal year 2014, the city established the North Nobiru Hill Promotion Council with the aim of building consensus among residents regarding the management and operation of the high-ground reclamation project.[16] As of fiscal year 2016, two councils coexist as active organizations in the Nobiru district: the Town Reconstruction Council, which represents the interests of current inhabitants of the district, and the North Nobiru Hill Promotion Council, which represents the interests of those wishing to move to the high-ground development district.

Yuriage District, Natori City

The reconstruction plans for the Yuriage district of Natori City feature land readjustment projects that provide a foundation for housing, commerce, and industry, while conducting disaster prevention measures (foundation-raising) for existing residential districts that were devastated by the tsunami. The local government organized the Natori City New Future Committee (hereafter Future Committee), which met from May to August 2011, with the goal of formulating reconstruction plans. However, the local government mismanaged the selection process of delegates for that meeting, and the process of formulating reconstruction plan proposals suffered accordingly.

When selecting members for the Future Committee, the local government called for participation by representatives from pre-disaster community organizations. However, the standards for choosing delegates were vague. For example, of the 11 *Chounaikais* in the Yuriage district, only two *Chounaikai* heads were included on the Future Committee.

Furthermore, in formulating reconstruction plans, the local government decided on land readjustment projects in existing residential districts by majority rule of the Future Committee members, without reviewing reconstruction plan proposals that took residents into consideration. After the disaster, Yuriage district residents launched several reconstruction council organizations intending to formulate their own reconstruction plans. Some of these reconstruction council organizations considered relocating as a group to inland areas and made such proposals to the local government. However, the local government did not adopt these proposals in their reconstruction plans.

In the latter half of 2011, several reconstruction council organizations sent declarations of intent to the local government about the reconstruction plan formulation procedures and decision details that the local government had made in the first half of that year. Residents ceaselessly criticized the Future Committee for "deciding everything with only some of the people." Yuriage residents were divided into "collective relocation" and "current site rebuilding" camps, and petitions and appeals to the city council and Miyagi Prefecture by reconstruction council organizations continued into fiscal year 2014.

In fiscal year 2014, the local government received permission from the national government concerning land readjustment projects for existing residential districts and began work on the reconstruction projects. It received advisors from Kobe City[17], which had been receiving guidance and support, and established the Yuriage District Town Reconstruction Council in May 2014, with the aim of building consensus among residents for land readjustment projects. Before founding the council, the local government held a session to exchange opinions in order to review residents' ideas about the selection of representatives and the management policies of the council.

Facilitator: What kinds of people would you like to become members of this council?

Is there anyone you would like to have lead it?

For a while, no one raises their hand to offer an opinion.

Resident A: How many Gyousei-Ku presidents are returning to Yuriage?

Facilitator: Are you saying that it would be best to have a Gyousei-Ku president lead the council?

Resident B: I don't think we need to be particular about that. We can freely participate without involvement from the city's existing neighborhood associations.

For me, I don't think it's necessary to emphasize official positions.

Dialogue for the Establishment of the Yuriage Town Reconstruction Council [February 21, 2014]

As a result of the dialogue, the local government and Yuriage residents decided that they would position the Yuriage Town Reconstruction Council as the sole point of contact for determining "resident consensus of opinion" on issues related to land readjustment projects and that council members would be solicited by open recruitment. Since fiscal year 2014, the Yuriage Town Reconstruction Council has been making proposals in cooperation with the local government concerning public housing, road configuration, and attracting business facilities to the area.

SUMMARY AND CONCLUSIONS

The findings of this study can be summarized in three points: First, citizen participation in the reconstruction process in areas struck by the Great East Japan Earthquake disaster was carried out on a foundation of community organizations. Second, traditional community organizations that had been founded in the area before the disaster contributed to reconciling residents' interests in the early stages of reconstruction. Third, local governments and community organizations drove changes to the frameworks of citizen participation from earlier stages, in proportion to time passed since the disaster and the progress of reconstruction projects.

In the course of disaster reconstruction, community organizations become the bases for citizen participation. However, unless local governments can regulate the activities and roles of community organizations through public institutions and policy procedures, citizen representation in the community organizations will not be guaranteed. As a result, in local governments that cannot effectively establish the legitimacy of policies, consensus-building fails and regional disparities in the pace of reconstruction ensue.

Previous research identified that, in the disaster reconstruction process, traditional community organizations acted as intermediaries between national and local governments and residents and that they became the foundations for citizen participation (Daly, 2014; Husin & Alvisyahrin, 2014). Certainly, in the cases studied in this chapter, the industry associations of Onagawa and the *Gyousei-Ku* of Higashimatsushima and other traditional community organizations that were formed before the disaster became foundations for citizen participation. However, in regional communities after the disaster, residents who had not belonged to traditional community organizations before the disaster had new incentives to participate in reconstruction projects—and indeed in the rebuilding of their lives. Therefore, in order for citizen organizations and movements that emerge after disaster to participate in policy processes, local governments and community organizations must institutionalize and reorganize. Taking into account the idiosyncrasies of prior research cases, traditional community organizations are shown to become the foundation for citizen participation in the long-term reconstruction process, and it is speculated that this is limited to areas in which urbanization and industrialization have not progressed.

In the case studies of this chapter, it has become clear that the existence of pre-disaster municipal mergers was not unrelated to governance in the disaster reconstruction process in disaster-struck areas (Cho, 2014). In Higashimatsushima City, where reconstruction progressed quickly, the municipal merger was an opportunity to institutionalize community organizations, which functioned as places to reconcile citizens' interests in the disaster reconstruction process. However, in Onagawa Town and Natori City, neither of which had experienced a municipal merger, pre-disaster community organization preparation and citizen participation levels differed. Therefore, Cho's thesis, that decentralization reforms and municipal mergers result in social and spatial inequalities during reconstruction, needs to be reexamined in light of local governmental citizen participation policies and the realities of citizen participation in local communities.

NOTES

1. In 1999, the Japanese government formulated the Decentralization Laws and revised Bill 475 related to the Local Autonomy Act. These reforms eliminated the administrative functions that the central government had been imposing on local governments, thus strengthening the policy-making authority of local governments.

2. From 2011 to 2015, the Japanese government allocated a budget of ¥395,300,000 for reconstruction subsidies. By using subsidies, local governments had no substantial financial burden in implementing reconstruction projects, and the reconstruction budget was 100% at the national expense (Kitamura, 2015).

3. The Japanese government recommends that local governments implement residential infrastructure renovation projects accompanying relocation efforts, such as disaster prevention collective relocation promotion projects and post-disaster public housing renovation projects.

4. The concept of citizen participation as applied in this study encompasses not only institutions established around the time of the disaster but also political mechanisms for citizen participation by local governments and local communities, including neighborhood movements that pop up in the regional community in the process of disaster reconstruction.

5. Daly, as well as Husin and Alvisyahrin, referred to the role that Gampong and Gotong Royong, traditional community organizations in the Indonesian state of Aceh, played in saving lives and distributing relief goods in areas affected by the Great Sumatra Earthquake of 2004.

6. The Onagawa Nuclear Power Plant started operation in Onagawa Town in 1984. The Japan Air Self-Defense Force, Matsushima base was placed in Higashimatsushima City before World War II, and Sendai Airport was opened in 1998, upgrading the prewar Natori Airfield.

7. As of 1957, there were eight fishery associations in the town of Onagawa, with a total union membership of 1076 (Onagawa Town, 1960). However, in 1993 the fishery associations were merged into one, and as of 2007, union membership was down to 621 people (Miyagi Prefectural Government Agriculture, Forestry and Fisheries Department, 2008).

8. Since the 1983 mayoral election, the main electoral issue was distribution of funds from the nuclear power plant. Of the seven elections between the 1983 election

and the pre-disaster 2007 election, three had election campaigns, and four were won by default (Kuroda & Tsuji, forthcoming).

9. Referring to documents regarding nuclear power plant counter-measures, provided by Onagawa Town.

10. Referring to a city's ordinance "Higashimatsushima City Gyousei-Ku Rules for Establishing Leadership."

11. From the Chamber of Commerce and Industry Yuriage Revitalization Committee, established in 2006, and others.

12. Onagawa Town website "Post-Earthquake Town Reconstruction Dialogue" http://www.town.onagawa.miyagi.jp/hukkou/kakusyu_setumeikai.html (June 20, 2016).

13. Higashimatsushima enacted the "Construction Regulations for Tsunami Disaster Prevention Areas" on March 15, 2012. Tsunami disaster prevention areas are classified into three types, and districts in which homes were spared from tsunami destruction fall under the third category. In third category districts, residents may rebuild or repair homes by raising the foundation 1.5 m above the ground.

14. The Nobiru District Zaitaku Residents Association was reorganized on May 31, 2012, into the Nobiru/Miyado District Association for Considering Restoration and Reconstruction.

15. The details of what the Zaitaku Residents Association demanded of the local government include provision of evacuation roads, attracting business facilities, etc.

16. The details of what the North Nobiru Hill Promotion Council considered are wide ranging, from decisions about housing block assignments to residence rules for the residential development districts.

17. Kobe city government has sent land readjustment project managers to Natori for long-term stationing there.

ACKNOWLEDGMENTS

This work was supported by Grant-in-Aid for JSPS Research Fellow. I would like to offer my special thanks to all cooperators in my fieldwork sites.

REFERENCES

Ahrens, J. & Rudolph, P. M. (2006). The importance of governance in risk reduction and disaster management. *Journal of Contingencies and Crisis Management*, *14*(4), 207–220.

Asahi Shimbun. (1995). Sendai-shi tono Gappei ha Fuseiritsu, Natori-shigikai, Gappeikyogikai Secchi Hiketsu de ['Rejection of the Natori City's Merger with Sendai City: Natori City Assembly Voted Down the Proposal of Establish the Council for Merger'] (in Japanese), 26 August.

Asahi Shimbun. (1996). 7 gatsu no Natori Shichosen, Gappei Souten, Ugoki Kappatsuni: 2Jinei Jimushobiraki [Natori City Mayoral Election will Take Place with the Issues of Municipal Merger in July: Both Camps have Opened Election Office] (in Japanese), 19 May.

Berke, P. R., Kartes, J., & Wenger, D. (1993). Recovery after disaster: Achieving sustainable development, mitigation and equity. *Disasters*, *17*(2), 93–109.

Cho, A. (2014). Post-tsunami recovery and reconstruction: Governance issues and implications of the Great East Japan Earthquake. *DISASTERS*, *38*(2), 157–178.

Daly, P. (2014). Embedded wisdom or rooted problems? Aid workers' perspectives on local social and political infrastructure in Posttsunami Aceh. *DISASTERS*, *39*(2), 232–257.

Davis, D. E. (2005). Reverberations: Mexico City's 1985 Earthquake and the transformation of the capital. In J. V. Lawrence, & T. J. Campanella (Eds.), *The Resilient City: How modern cities recover from disaster* (pp. 255–280). Oxford: Oxford University Press.

Edgington, D. W. (2010). *Reconstructing KOBE: The geography of crisis and opportunity*. Vancouver: UBC Press.

Güiza, F., Simmons, P., Burgess, J., & McCall, M. K. (2016). Chronic institutional failure and enhanced vulnerability to flash-floods in the Cuenca Altadel Río Lerma, Mexico. *DISASTERS*, *40*(1), 112–133.

Hilhorst, D. (2013). *Disaster, conflict and society in disaster: Everyday politics of crisis response*. New York: Routledge.

Husin, T., & Alvisyahrin, T. (2014). Role of community and communal law of Aceh in the Great Sumatra Earthquake and Tsunami recovery: A case study in Lambada Lhok Village, Aceh Besar District, Aceh, Indonesia. *Journal of International Cooperation Studies, Kobe University*, *21*, 63–79.

Iwasaki, N. (1998). 'Kokka-toshi' Kobe no Higeki to 'Shimin Shakai' no Kuto: Hanshin-Awaji Daishinsai kara Mietekurukoto [The Tragedy of 'State City' Kobe and the Struggle of 'Civil Society': What We Can See in the Disaster of the Great Hanshin Earthquake] (in Japanese), *Annals of Regional and Community Studies*, *10*, 1–18.

Ishinomaki Kahoku. (1983). Genpatsu Meguri Machi Nibun, Sanagara 'Jumin Touhyo' ['Residents of Onagawa Town Split into Two Camps over Nuclear Power Plant: Mayoral Election was just like a Referendum] (in Japanese), 27 January.

Kahoku Shimpo. (2003). Oshika-cho gikai, Onagawa-cho tono Gappei Tekkai wo, Ishinomaki chiku heno gouryu wo youbou [Oshika Town Assembly may withdraw the Municipal Merger with Onagawa Town: They will Demand the Merger with Municipalities in Ishinomaki Region] (in Japanese), 3 April.

Katoh, H. (2015). March 11. (Higashinihon-Daishinsai 4nen) Saiken Konosakiha, Jyutaku Tetsudou Douro Genpatsu hinan [4 Years After the Great East Japan Earthquake: What will Become of Reconstruction, Housing, Railroads, Roads and Residents Evacuation from the Nuclear Power Plant] (in Japanese), *Asahi Shimbun*.

Kitamura, W. (2015). Hisaijichitai ni taisuru Seifu no Zaiseisochi [Government Financial Measures towards Local Governments around the disaster affected areas] (in Japanese). In T. Kohara & H. Inatsugu (Eds.), *Shinsaigo no Jichitai Governance [Local governance after disaster]* (pp. 121–145). Tokyo: TOYO KEIZAI INC.

Kuroda, Y. & Tsuji, T. (forthcoming). Onagawa-cho no Fukko to Genpatsu: Genpatsu to Chiiki Shakai [Disaster Reconstruction and Nuclear Power Plant in Onagawa Town: Local Community and Nuclear Power Plant] (in Japanese). In H. Yoshino, M. Kato, & Y. Yamashita (Eds.), *Shinsai to Tohoku: Atarasii Tohoku wo Mezashite* [Disaster and Tohoku Region: Aiming at New Tohoku]. Tokyo: YUHIKAKU.

Lidskog, R., & Sjödin, D. (2015). Time, knowledge, and risk: Decision making. In the aftermath of storm disasters. *International Journal of Mass Emergencies and Disasters*, *33*(3), 341–359.

Lombard, M. (2013). Citizen participation in urban governance in the context of democratization: Evidence from low-income neighborhoods in Mexico. *International Journal of Urban and Regional Research*, *37*(1), 135–150.

Miyagi Prefectural Government Agriculture, Forestry and Fisheries Department. (2008). *Miyagiken Suisangyou Kyoudou Kumiai Nenpou, Heisei 19 Nendo Ban [Annual Report of Miyagi Prefectural Fisheries Cooperative Association, Fiscal 2008 Ver.]* (in Japanese). Sendai: Miyagi Prefectural Government.

Naruse Town News. (2002). 4gatsu 1nichi kara, Minasan no Koe Shokuin ga Chouchou ni Todokemasu, Machizukuri Chiku Tantouin tou Seido Start ['We are going to Deliver Your Opinions to Mayor: the Launch of Personnel Assignment System for Community Development, April 1] (in Japanese). April.

Natori City. (1977). *Natori-shi shi [History of Natori City]* (in Japanese). Natori: Natori City.

Natori Commerce and Industry News. (2001). Chiku Dayori [News from each district] (in Japanese). 1 January.

Nozaki, R. (2005). Machi-zukuri Kadai no Keisho to Hatten [Inheritance and development Relating to Future Community Development Issues] (in Japanese). In Shinsai 10nen Shimin Kenshou Iinkai (Ed.), *Hanshin-Awaji Daishinsai 10nen [10 Years After the Great Hanshin-Awaji Earthquake]* (pp. 49–60). Kyoto: BUNRIKAKU.

Olson, R. S. (2000). Toward a politics of disaster: Losses, values, agendas, and blame. *International Journal of Mass Emergencies and Disasters*, *18*(2), 265–287.

Onagawa Town. (1960). *Onagawa Choushi [History of Onagawa Town]* (in Japanese). Onagawa: Onagawa Town.

Onagawa Town News. (1992). Heisei 4nendo, Mizuage Yushusen narabini Yuryo Kaiukenin Hyoushoshiki kara [Awards Ceremony of the Most Valuable Landing Ships and Purchasers, Fiscal 2004] (in Japanese). 1 August.

Okazaki, I. (1977). *Yuriage Fudoki [Topography of Yuriage district]* (in Japanese). Natori: Ono shinpei.

Sakurai, I. (2015). December 30. 2015 News Kaisou, Shuzai Note kara (8), Higashimatsushima-shi de Shin Shigaichi Zokuzoku Tanjyo ['Looking back on 2015 from Interview notes: Newly Developed Areas have opened one after another in Higashimatsushima City'] (in Japanese). *Ishinomaki Kahoku.*

Satoh, K. (2007). Juumin Sanka Kenkyu no Rironteki Shiza: Governance Ron no Shiten Kara ['A Theoretical Perspective on Citizen Participation: From the Viewpoint of Governance Theory'] (in Japanese). *Studies of regional policy, Takasaki City University of Economics*, *10*(2), 45–61.

Smith, G., & Birkland, T. (2012). Building a theory of recovery: Institutional dimensions. *International Journal of Mass Emergencies and Disasters*, *30*(2), 147–170.

Tierney, K. (2012). Disaster governance: Social, political, and economic dimensions. *Annual Review of Environment and Resources*, *37*, 341–363.

Tierney, K. (2014). *The social roots of risk: Producing disasters, promoting resilience*. California: Stanford University Press.

Watanabe, T. (1999). *'Onagawa Genpatsu' Chiiki to Tomoni ['Onagawa Nuclear Power Plant with Local Community']* (in Japanese). Tokyo: TOYO KEIZAI INC.

CHAPTER 7

SOCIAL VULNERABILITY IN DISASTERS: IMMIGRANT AND REFUGEE EXPERIENCES IN CANTERBURY AND TOHOKU

Shinya Uekusa

ABSTRACT

This comparative study qualitatively explores how linguistic minority immigrants and refugees experienced the 2010–2011 Canterbury and Tohoku disasters, including their coping mechanisms and their perceived vulnerabilities and resilience. The data used for this qualitative analysis was primarily drawn from 28 in-depth interviews with linguistic minority immigrants and refugees and their supporting organization staff conducted in 2015–2016. Additional material was drawn from two publicly available data sets. Immigrants and refugees are typically thought of as being more vulnerable in disasters. However, findings drawn from this research demonstrate the nonlinearity, complexity, and contextuality of social vulnerabilities in disasters, suggesting that they are not necessarily powerless help-seekers in some cases. Using Bourdieu's capital theory, this study demonstrates how immigrants and refugees were active social

Recovering from Catastrophic Disaster in Asia
Community, Environment and Disaster Risk Management, Volume 18, 127–144

ISSN: 2040-7262/doi:10.1108/S2040-726220160000018006

agents in these disasters. Consequently, we need to reconceptualize the social vulnerability approach. Some study participants had experiences of going through wars and "everyday disasters," which made them more "resilient." This is conceptualized here as "earned strength," which can be a significant resource in disasters for the socially vulnerable. This chapter hopes to answer some critical questions regarding the social vulnerability approach: how do we incorporate the structure–agency concept, how do we theoretically deal with the contextuality/nonlinearity of social vulnerability in disasters, and how do we conceptualize a research study that can seek more practical and generalizable findings, instead of event-driven and disaster-specific findings?

Keywords: Capital; earned strength; everyday disasters; Pierre Bourdieu; resilience; social vulnerability

INTRODUCTION

This comparative study qualitatively explores how immigrants and refugees who are linguistic minorities in Canterbury, New Zealand, and Tohoku, Japan, experienced the 2010–2011 disasters, including their coping mechanisms and their perceived social vulnerabilities. Sociological research on disasters has found that disasters affect individuals, families, and communities differently, and damage is usually uneven owing to the structural inequalities that already exist prior to disasters (e.g., Matthewman, 2015; Tierney, 2014). Immigrants, refugees, and linguistic minorities are typically considered more vulnerable in disasters than the general population. However, findings drawn from in-depth interviews demonstrate the nonlinearity, complexity, and contextuality of social vulnerabilities in disasters, possibly suggesting that these people are not necessarily powerless help-seekers in some cases and that we, therefore, need to reconceptualize the social vulnerability approach. Using Bourdieu's capital theory (1986), this study demonstrates how linguistic minority immigrants and refugees were active social agents in these disasters. The ways in which they individually and collectively generated, deployed, and employed a variety of resources/capital to adapt to the rapidly changing situations can offer practical theoretical and policy implications.

BRIEF REVIEW OF SOCIAL VULNERABILITY LITERATURE

Unlike traditional disaster research approaches that focus on the environmental, spatial, and physical factors (such as building structure and where people live), disaster sociologists emphasize the social factors that contribute to uneven disaster damages and investigate how those social factors affect individuals, families, and communities differently in disasters. Numerous empirical studies (e.g., Aldrich, 2012a; Enarson, 2007; Lindell, 2013; Peacock, Morrow, & Gladwin, 1997; Tierney, 2014; Webb, 2007) have confirmed that it is not only disaster damage but also all phases and aspects of disaster such as disaster risks, disaster preparedness, disaster recovery, and disaster resilience. These factors are influenced by social categories such as class, race, and sex. As a result, some individuals and groups are affected more severely during disasters, and, after disasters, they receive less support/resources than others, resulting in slower or limited recovery (see, e.g., Tierney, 2014).

Aldrich (2012b) explains that "[s]ociologists and economists have focused on the socioeconomic status of victims and have tried to link their recovery processes to their wealth and private resources" (p. 161). Hurricane Katrina was a critical event that seemed to clearly demonstrate such a link: the resulting damage was uneven such that some individuals and communities, particularly racial minorities and low-income communities, suffered more harm and experienced slower or limited recovery. Similarly, Aldrich (2012b) found that "[m]any of the communities struck hardest by the 2011 Tohoku earthquake and tsunami were older, retired residents with little savings and no home or earthquake insurance" (p. 161). These studies, along with numerous others, have confirmed that the socially vulnerable such as racial/ethnic minorities, the poor, and elderly seem to suffer more from disasters and risks, experiencing greater disaster damage and poorer recovery, or they reside in more disaster prone areas (e.g., Cutter, Boruff, & Shirley, 2003; Matthewman, 2015; Tierney, 2014).

The social vulnerability approach has been a very popular way to explain differential suffering in disasters. As such it has played a critical role in a paradigm shift in disaster studies. Scholars now theorize that the socially vulnerable, who are more likely to suffer during and after disasters, are often *systematically* put in high disaster risk (and environmental hazard) situations (see, e.g., Matthewman, 2015). Bankoff (2006) explains that "[s]ocial systems generate unequal exposure to risk by making some people more prone to disaster than others and these inequalities are largely a function of the power relations (class, age, gender, and ethnicity among others) operative in every society" (quoted in Fordham, Lovekamp, Thomas, & Philips, 2013, p. 5). Indeed, social scientists have conducted numerous empirical studies to explore how these social/demographic factors generate social vulnerability in disasters and systematically disadvantage certain social groups (e.g., Cutter et al., 2003; Enarson, 2007; Morrow, 1999). This paradigm shift, largely influenced by feminist disaster scholars and sociologists, has been extremely helpful in theorizing and identifying where the vulnerable populations are in disasters and making them more visible to researchers, practitioners, and policymakers. For example, Cutter et al. (2003) developed the Social Vulnerability Index (SoVI), which may be one of the most comprehensive and a useful theoretical models to help researchers, practitioners, and policymakers identify socially vulnerable groups in a given geographic area.

Some Contradictions: Are the Socially Disadvantaged Necessarily Vulnerable to Disasters?

Nonetheless, socially vulnerable individuals and communities may not necessarily be powerless victims in disasters and may show *unexpectedly* quicker recovery and stronger resilience than their counterparts, just like the Vietnamese community in Eastern New Orleans after Hurricane Katrina (Leong et al., 2007), Māori communities in Christchurch after the 2010–2011 earthquakes (Kenney & Phibbs 2015; Lambert, 2014), Latino immigrant farmworkers after the 1994 Northridge earthquake (Bolin & Stanford, 1998), Latinos during the 1995 Chicago heat wave (Klinenberg, 2003), and the Filipina community in the 2011 Tohoku disaster (Lee, 2012). There are, indeed, many examples found in the current literature, which contradict our common understanding of social vulnerability in disasters.

Are these success stories simply exceptions to the general rule of social vulnerability? These examples are treated as "exceptions" in some cases, while Bolin and Stanford (1998), Enarson (2007), and other scholars suggest that social solidarity may be a more important factor than traditional social factors such as class when it comes to coping with disasters. Peacock, Van Zandt, Zhang, and Highfield (2015) argue that "[o]ur panel model suggests that social vulnerability issues are important in the recovery process; however, the nature of their influence may be more nuanced than the literature would suggest" (p. 363). A major shortcoming of the traditional social vulnerability approach is its excessive emphasis on social structure. This means that it fails to sufficiently consider social agency. People in disasters may not be as powerless as many assume. Therefore, the nonlinearity, complexity, and contextuality of social vulnerability in disasters should be more thoroughly addressed in disaster research.

Critical questions for disaster researchers become apparent here: how do we incorporate the structure–agency concept? How do we theoretically deal with the contextuality and nonlinearity of social vulnerability in disasters? How do we conceptualize a research study which can seek more practical and generalizable findings, instead of event-driven and disaster-specific findings? This chapter hopes to answer these questions, employing Bourdieu's theoretical concepts in addition to intersectionality and feminist perspectives, all of which have contributed to more nuanced understanding of disasters across the last few decades.

THEORETICAL PERSPECTIVES

Bourdieu's theory of capital (Bourdieu, 1986) provides a useful conceptual framework for disaster research. Bourdieu (1986) uses capital to refer to resources in the broadest sense (p. 243). This capital may be economic (financial assets), cultural (skills, attitudes, knowledge, language, norms, experience, and education), social (cohesion, networks, and group membership), or symbolic (rewards accruing from status, respectability, and reputation). Capital possession determines one's place in the social order, so Bourdieu's work is normally used to explain inequality and its perpetuation. However, some of these noneconomic forms of capital, particularly social capital, can help disaster researchers understand how groups can deal with social vulnerability in disasters. Indeed, social capital theory has been popular in disaster studies, which emphasize resilience and social agency (Aldrich, 2012a; Tierney, 2014). However, these studies have *partially* employed Bourdieu's capital theory, because Bourdieu also emphasizes the totality and interaction of capitals. This means that the quality of, for example, social capital may depend on *other* forms of capital(s), because capital may not operate independently and can convert into other form(s) of capital (Bourdieu, 1986).

Bourdieu's capital is not normally found in disaster scholarship; however, his concepts can provide a useful way to examine how the quantity, quality, and interplay of capitals affect the experiences of individuals and groups differently in disasters. His theoretical perspectives help researchers assume their social agency and hypothesize that social agents are actively involved in creating social reality through the dialectical relationship between structure and agency (or habitus and field), always responding to constantly and rapidly changing social situations in disaster contexts. His emphasis on the dialectic between structure and agency, in particular, can help disaster researchers overcome the typical structure–agency and theory-practice dilemma because, as Enarson, Forhergill, and Peek (2007) note, most disaster studies have been strong in practice/description but struggle with theory, focusing on impacts and responses but not necessarily underlying issues and social agency (p. 139). His theoretical notions are, therefore, helpful for disaster researchers as they reconceptualize the traditional macrostructural social vulnerability approach, assuming that individuals are active social agents, instead of powerless victims.

METHODOLOGY AND DATA COLLECTION

The data used for my qualitative analysis were primarily drawn from 28 in-depth interviews with immigrants, refugees, linguistic minorities, and their supporting organization staff conducted in Canterbury, New Zealand, and Tohoku, Japan, in 2015 and 2016. Qualitative methods allowed me to examine and explore the concrete ways in which some of the socially vulnerable have experienced, coped with, and negotiated disasters. Phillips (2014) argues that "[b]y designing an inductive approach that directs researchers to create explanations from the data (rather than deductively based hypothesis-testing), it is possible to capture, analyze and report on social phenomena surrounding unanticipated events" (p. 3). To explore the experiences of the socially vulnerable in more authentic ways, qualitative interviews with those who experienced the disasters have been utilized, because although it is ideal, most researchers are unable to conduct participatory observations (except those who happened to be in the disaster situations with the study populations or can physically be in the communities during the recovery process). It is imperative that disaster relief and management be informed by the experiences of those who have lived through them.

I conducted 28 in-depth interviews with people who experienced the 2010–2011 disasters in Canterbury or Tohoku (15 in Canterbury and 13 in Tohoku). These interviewees include 13 males and 15 females. They were of various age groups (between 20s and 50s). They were of diverse nationalities and cultural backgrounds (such as Afghan, Argentinean, Chinese, Filipina, German, Indian, Iraqi, Japanese, Korean, Pacific Island, Russian, Scottish, Somali, and South African). In doing so, this study hopes to explore some common features of these diverse groups and to capture a wide range of disaster experiences.

They were contacted through the researcher's personal networks, colleagues, and supporting organizations. Then more interviewees were selected through these first sets of interviewees, using snowball sampling. All interviews were informal and open-ended. They were conducted at locations comfortable for the interviewees. A set of open-ended interview questions were created prior to the interviews on the basis of my pilot study and literature review. All interviews were audio recorded, and I later transcribed (and translated) them for qualitative analysis. Pseudonyms are used in all publications. Some publicly available data such as "Women's Voices" (Christchurch Branch of National Council of Women of NZ, 2014) and *Experiencing Disasters in a Foreign Land: 3/11 for Koreans Living in Tohoku* (The Great East Japan Disasters Korean Immigrants Interview Project, 2015) was also used to include more voices of linguistic minority immigrants and refugees in these disaster-affected areas.

FINDINGS

Interviewees noted a variety of strategies and capital that they used to cope with the disasters and their aftermath. The ways in which these interviewees have experienced, coped, and negotiated with the disasters and recovery confirmed previous research findings (e.g., Kenney & Phibbs, 2015; Lee, 2012; Leong et al., 2007; Marlowe & Lou, 2013). Immigrants, refugees, and linguistic minorities are extremely vulnerable in disasters due to the structural inequalities that exist prior to disasters, yet some of them are remarkably resilient because of their social agency and life experiences. Cross-cultural findings in this chapter demonstrate that social vulnerability in disasters is not static but rather nonlinear and contextual.

Structural Constraints and Social Vulnerability

Structural constraints have confronted immigrants, refugees, and linguistic minorities in these disasters. Indeed, regardless of location, most study informants' experiences confirmed their social vulnerabilities to disasters. For example, language barriers, institutionalized "linguicism" (see Skutnabb-Kangas & Phillipson [1995] for linguicism and language-based discrimination) or lack of linguistic capital was one of the most noticeable perceived disadvantages that many of the study informants needed to deal with in the disasters in both locations. One of the "Women's Voices" interviewees, Hana, a Korean migrant woman in Canterbury, told about her linguistic and psychological struggle of dealing with the insurance company:

> I don't want to argue with my insurance company. I felt kind of disadvantaged with my lack of English skills, so my family decided the easy option.... We accept [the] Government's offer, Option 1 [instead of negotiating a settlement on our house with the insurance company]. I don't have the confidence with the language, so I gave up. (p. 5)

Settlement with insurance companies is often time consuming and difficult even for native speakers, yet, in this excerpt, the perceived linguistic disadvantage, which native speakers normally do not confront, particularly after disasters, is quite obvious.

My Tohoku migrant interviewees, namely Maria (a Filipina migrant woman) and Ji-yoon (a Korean migrant woman), explained more direct forms of linguistic barriers and linguicism that they experienced during the Tohoku disaster: critical information such as tsunami warnings and evacuation information was simply not accessible, particularly for the newcomers who were not competent in Japanese. For instance, Maria told a story about a Filipina migrant woman:

> one of the saddest things was, one of us perished in [the] tsunami.... Well, the thing is that she must have not understood the [tsunami warning] announcement, because the announcement was in Japanese, you know? And the announcement was using these words that are not being used in everyday life.

Maria's friend's case may be extreme, yet previous research provides overwhelming evidence that language barriers are a typical structural disadvantage that migrants and refugees, who are linguistic minorities in any given society, often have to confront in disasters (e.g., Santos-Hernández & Morrow, 2013).

However, the issue may not just be the difficulty of understanding the key disaster terms and critical information in dominant languages; rather, as Maria and Belaynesh in "Women's Voices" (an Ethiopian refugee woman in Canterbury) stated, they needed someone to talk to and express their *real* feelings to in their languages during the frightening and stressful times of disaster and recovery. Yet, as Hana, Ji-yoon, Belaynesh, and Nadifa (a Somali refugee woman in Canterbury) all noted, linguicism was not something that suddenly emerged during the disasters; it is a consistent form of social discrimination which exists in their everyday life in the host countries. It might have only become *visible* to the public in these disasters, but, for them, it was nothing new.

Social Capital, Communitas, and Social Agency

The standard social science expectation is that the poor will suffer most in disasters; consequently, most of the social sciences' focus has been on victims' socio-economic standing (Aldrich, 2012b). Yet, in some disasters, poor groups may cope better than others (Klinenberg, 2003). This suggests that there can be important resources beyond the merely financial, and social vulnerability may be more contextual and nonlinear than is ordinarily supposed. For example, Klinenberg's (2003) study stresses the value of what some disaster researchers might term "social infrastructure," the development of neighborhood ecologies of support. Bourdieu would understand this as social capital. Indeed, social capital has been a popular concept that disaster researchers such as Aldrich (2012a) and Tierney (2014) have used to examine and explore social agency and disaster resilience among the socially disadvantaged. Like many other empirical studies, my findings provide similar evidence that social capital was (or brought) critical resources, which enabled the typically socially disadvantaged migrants, refugees, and linguistic minorities to get by in these disasters.

Just like the Vietnamese Catholic community in Eastern New Orleans after Hurricane Katrina (Leong et al., 2007), the study informants described that religion was one of the most significant resources, which gave them positive attitudes and encouragement to psychologically cope with the disasters (as cultural capital) and enabled them to develop and expand their already-durable social networks (as social capital) to obtain logistic, material, financial, and emotional support. As informed by religious group leaders in my New Zealand and Japan interviews, volunteers, and donations flowed into their local churches from all over the world through their already-existing and improvisationally developed social networks and then redistributed to the preexisting local religious communities. Indeed, these study informants' experiences confirmed the importance of social capital and demonstrated the way in which they purposefully and extemporaneously developed and employed their extended and transnational social networks to deal with the material shortage and disaster difficulties.

However, it is worth stressing here, especially for future research, that social capital may not be something that they strategically developed during disasters. Socially isolated and marginalized communities, particularly racial minorities, refugees, immigrants, and linguistic minorities, may develop a durable social network because they often lack other forms of capital, particularly economic and cultural, and tend to rely on their social capital as their main source of guidance and help in competitive social fields such as labor markets and education (Morrow, 2008). It is usually a common feature of socially marginalized communities because it can be an outcome of social marginalization and linguistic/cultural isolation in any given society. Hence, their preexisting durable social networks may not be a part of their disaster preparedness.

Furthermore, many interviewees mentioned the generous help that they received from their neighbors or random people. A "Women's Voices" project interviewer summarized the interview with Belaynesh and reported that

> [Belaynesh's] neighbour came to the door to check up on [Belaynesh]. At first Belaynesh did not want to open the door, but the neighbour said who he was and, in the end, she opened the door. It was good to see her neighbour who had a torch. He checked to see that she and her children were OK. "He makes us settle down.... My head is no good... Really hard and very scary.... I find it difficult" (p. 4).

As we can see here, one of the reasons why these interviewees were able to get by in the disasters was the "extraordinary community" in Solnit's term (2009) or *communitas* as Turner (1969) defines it, which improvisationally arises during disasters. According to Drury, Cocking, and Reacher (2009), Matthewman, (2015), Solnit (2009), and Tierney (2009), when a calamitous event occurs, individuals tend to help each other, instead of being selfish, thereby creating improvisational and possibly temporal cultural capital to self-organize and extend their social networks to help others, often trying to even reach out to the invisible socially marginalized communities. *Communitas* may be understood as significant cultural capital, which appears as society-wide attitudes and belief to help each other and deal with the disaster damage and loss collectively, often resulting in extemporaneous social networks and enhancing the already existing social networks to better deal with the disasters. The key here is that social agents organically, but not strategically, create a form of cultural capital in response to the environmental and contextual changes, recognizing the *need* to help each other and the powerless victims.

Many disaster researchers have argued that some socially vulnerable individuals and communities have shown remarkable recovery, adaptability, and copability to disasters because they are resilient. The typical logic being assumed here by many disaster researchers is that resilience can reduce vulnerability. This has become central to social agency based disaster theories but is somehow misleading as vulnerable individuals and communities *can* become resilient, actively coping with disaster damage and adapting to the new normal (Uekusa & Matthewman, 2017). However, some of the experiences that these study informants described were somewhat nonaligned because some of them *involuntarily* become resilient. In the following section, I will discuss the paradox of social vulnerability and resilience based on the unexamined theme of "earned strength," which emerged during my analysis.

Earned Strength

Some of the ways in which these study informants dealt with the disasters may not be deliberate. Unlike strategically developing and using social capital, their coping mechanism with these disasters was based more on their previous experiences. These made them strong enough to get by. As McIntosh (2007) notes, this can be called an example of earned strength, which can provide significant cultural capital in extreme events for the socially vulnerable to cope with disasters and adapt to the "new normal."

Some of my Canterbury interviewees and "Women's Voices" interviewees, such as Afghan, Iraqi, and Somali refugees, demonstrated how their previous experiences of going through wars, displacement, and then resettlement and social inequality in a foreign country meant that the series of Canterbury earthquakes was, while scary, still quite manageable. When I asked them why they and their communities were resilient and how they coped so well with the earthquakes, Nadifa explained that

> because... [the] Somali community... we had the Civil War, and we know how to be sturdy. We encourage our children, whatever happens, they have to be sturdy. They have to be successful because if they are sturdy, and if they are sturdy [at a] high level, whatever they go through, they will be successful.

Nadifa did not expand upon in any detail, yet she was aware that it was past hardships that had helped the Somali community in the Canterbury earthquakes. It is conceivable that their previous experiences *unexpectedly* helped them have (religiously based) strong social networks, cultural attitudes, disaster copability, and adaptability to the new challenges.

Similarly, another interviewee, Dari (an Afghan refugee woman in Canterbury), explained that

> ...refugees have gone through... all sorts of issues. Probably some of them could not cope and they did not want any more... you know, headache, emotional stress... so they left [after the earthquakes]. But a lot of us have stayed here as well, thinking, oh, well, at least, like... well, nobody is shooting at us. Yes, it's hard, yet we have been looked after, so we would stay, you know? At least, there is peace, there is support here. We are not kicked out of the city...

As Dari observed, the Canterbury earthquakes were further traumatic experiences for some of the Afghan refugees who could not deal with anymore and moved away. However, for Dari and her community, it was more manageable as compared with their previous frightening, traumatic, and life-threatening experiences, especially because they had support in Canterbury. Nadifa, Dari, and their communities were still socially disadvantaged but had enough earned strength to cope with the earthquakes and their aftermath.

Likewise, the other Canterbury interviewee, Jahmir (an Iraqi refugee male and former soldier), lived through multiple wars and had first-hand battlefield experience, which made the earthquakes minor traumas for him. He said that

> I thought, for me, [the earthquake] wasn't like too hard... my room was shaking, everything was shaking. I knew what was going on. I knew that was [an] earthquake. But, for me, it wasn't hard because I have been a soldier... I knew how to survive. [Since] I was 14 years old, my father taught me how to shoot the gun... I was a soldier and I have seen worse in the battlefields. [The earthquake] wasn't like that, you know? [In the battlefields,] I [was] walking 78 hours a week... I walk, no food, no water... my foot skin was all coming off... and I was carrying everything in my bag... so I knew how to survive after the earthquakes.

I then asked Jahmir if nothing was really difficult for him because of his past experiences. He answered

> Yea... no, it was nothing. I was scared [of the earthquake] but, no, [it was not difficult]. My kiwi landlord, she tried to get my family out of the house and said to them that I am brave. She went to check my son, but I was already carrying him out that time while the house was shaking, the ground was shaking, everything was shaking... my house was... the garage was collapsing.

His prior wartime experiences made it easier for him to deal with shaking ground, collapsing houses, distraught neighbors, and the post-quake chaos. Furthermore, in addition to refugees' experiences with "chaos" before coming to New Zealand like Jahmir's, because they restarted life in New Zealand without much capital, they had already collected and created capital such as developing durable social networks (church networks and ethnic communities) to depend on each other to collectively get by their everyday hardships.

These study informants had the significant cultural and social capital to be disaster *resilient*: prior experiences, practical knowledge, cultural values, and attitudes of how to support each other – in order to survive in chaotic situations, to survive without basic necessities such as water and power, and to *restart* their life without major government assistance. Therefore, those who face social inequality on an everyday basis might have "earned strength" to cope in disasters, because their everyday experiences of social marginalization contribute to their disaster resiliency (Davidson & Davidson, 2009). In this sense, earned strength acted as cultural capital. This was an important resource, an unintended consequence of everyday social inequality.

Similarly, Korean and Filipina immigrant women's stories from my Tohoku interviews and the book by The Great East Japan Disasters Korean Immigrants Interview Project (2015) further demonstrate how their earned strength as cultural capital helped them to deal with the disasters that impacted upon them. When I asked Maria why Filipina women were resilient, she answered

> I think it [has] something to do with our culture. We are not easily beaten by difficulties and hardships. Actually, the Philippines are not a progressive country, it is a poor country, we live in a difficult life, so that makes Filipinas stronger enough.

Like Nadifa's, her answer was quite abstract, not explicitly describing what exactly made them resilient, yet we can sociologically contextualize their experiences. In the first place, most of these immigrant women decided to move to Japan to get away from the everyday hardships and "bad luck" they had in their home countries. After being recruited, moving to Japan and getting married to Japanese husbands (most of them are farmers and fishermen), they now faced isolation, oppression, and discrimination due to the patriarchal host culture, the language barrier, the traditional Japanese family system, and different cultural norms and expectations. We can assume that their experiences of being oppressed both pre- and post-migration to Japan might have made them somewhat disaster resilient. It is worth noting that they were facing the actual Tohoku disaster *and* "everyday disasters" (Matthewman, 2015, p. 129–132) for being non-Japanese immigrant wives.

Yet earned strength may not apply to all socially marginalized groups or in some contexts. In the case of an Afghan refugee woman in Canterbury, the previous traumatic experiences made her less able to cope with the earthquakes and have nearly no resilience. Alia (an Afghan refugee woman), who went through the war in Afghanistan, dislocation, resettlement, and social isolation in New Zealand, explained that

> [seizure] happened after the after-shocks… I am still very scared of the earthquake, and tsunami. I think we can cope with the earthquake because the Civil Defense help and the children have special training, how to hold and cover, and go under the table, and go to a safe place – it's OK. Earthquake is no problem, it's just the tsunami, how should I do because in the past I have a very hard life. Just I want my next generation to be safe. I just worry about my children, my family. My family members, living in Afghanistan are not safe from Taliban, but here we have big step to move very far away from my country and this is the second tragedy – now we hope to be safe (p. 11).

Alia's psychological struggle in dealing with the earthquakes and aftershocks is obvious here. Indeed, Alia had a series of mental health issues and negative experiences such as insomnia and car accidents following the earthquakes. For Alia, the previous experiences are, on the contrary, traumatic and made her even more vulnerable in the disasters. Therefore, we should not misunderstand and misconceptualize earned strength, because previous negative experiences do not necessarily transform into positive cultural capital. Sometimes people's vulnerability is compounded, and there are limits of resilience.

Thus, the concept of social vulnerability and the relationship between vulnerability and resilience may be more complex than many of us have so far acknowledged. It should not be conceptualized as a static and either/or phenomenon, because vulnerability and resilience may not necessarily be binary opposites. It is quite possible that some socially vulnerable people can be resilient partly because they are coping with the everyday inequalities, which already confront them, and because of their previous experiences of disasters and chaotic events. The paradox of resilience is obvious here, and further research is recommended.

DISCUSSION

For disaster researchers, resilience is the concept to explain and further investigate issues relating to disaster recovery. As Tierney (2014) notes, resilience has been heavily emphasized in disaster research, becoming very popular among practitioners and policymakers. Nonetheless, the term is troubled, and this chapter has suggested that resilience itself is a concept in need of explanation. Disaster research has moved to resilience thinking without thoroughly understanding social vulnerability in disasters. We need to comprehend social vulnerability to disasters and its relationship with resilience more fully. We also need greater emphasis upon people's power. Social vulnerability should be understood as being a more complex, contextual, and fluid concept than is commonly supposed, possibly because, as briefly discussed, some socially vulnerable are resilient in some contexts and/or some are resilient ironically *because* they are socially vulnerable. Social vulnerability and resilience are typically conceptualized as binary opposites, or are said to exist in an inverse relationship to each other (Rodin & Garris, 2012), but their relationship may not be this simple. The paradox of vulnerability and resilience, therefore, requires further investigation (Uekusa & Matthewman, 2017). In making this recommendation, I suggest drawing on the theoretical repertoires of Pierre Bourdieu and Peggy McIntosh.

CONCLUSION

This study has demonstrated the nonlinearity, complexity, and contextuality of social vulnerabilities in disasters, suggesting that these people are active social agents rather than powerless help-seekers in some cases. Using Bourdieu's capital theory (1986), this study demonstrated how linguistic minority immigrants and refugees, as active social agents, were socially vulnerable in these disasters; they individually and collectively generated, deployed, and employed a variety of capitals, particularly cultural and social, to adapt to the rapidly changing situations in which they suddenly found themselves. Furthermore, some study participants had past experiences of going through wars, disasters, chaotic events, and everyday social injustices, which made them more resilient and disaster "prepared." As McIntosh (2007) notes, this can be called an example of earned strength, which can unexpectedly be a significant resource in extreme events for the socially vulnerable. Therefore, these types of cultural and social capital seem less *strategic*; it may be an unintended consequence of the everyday disasters such that the socially disadvantaged had already created these types of capital to negotiate their everyday hardships.

ACKNOWLEDGMENTS

The author acknowledges support from Dr. William L. Waugh and Dr. Ziqiang Han. The author would like to thank Dr. Steve Matthewman for his helpful suggestions. The author also would like to thank the participants of this study.

REFERENCES

Aldrich, D. P. (2012a). *Building resilience: Social capital in post-disaster recovery*. Chicago, IL: The University of Chicago Press.

Aldrich, D. P. (2012b). Social capital in post disaster recovery: Towards a resilient and compassionate East Asian community. In Y. Sawada & O. Sothea (Eds.), *Economic and welfare impacts of disasters in East Asia and policy responses* (pp. 157–178). ERIA Research Project Report 2011-8, Jakarta.

Bankoff, G. (2006). The tale of the three pigs: Taking another look at vulnerability in the light of the Indian Ocean tsunami and Hurricane Katrina. *Understanding Katrina*. Retrieved from http://understandingkatrina.ssrc.org/Bankoff/

Bolin, R., & Stanford, L. (1998). The Northridge earthquake: Community-based approaches to unmet recovery needs. *Disasters*, *22*(1), 21–38.

Bourdieu, P. (1986). The forms of capital. In J. G. Richardson (Ed.) *Handbook of theory for the sociology of education* (pp. 241–258). New York, NY: Greenwood Press.

Christchurch Branch of National Council of Women of NZ. (2014). Women's Voices: Recording women's experiences of the Canterbury earthquakes. *University of Canterbury Quakestudies*. Retrieved from https://quakestudies.canterbury.ac.nz/store/collection/228

Cutter, S. L., Boruff, B. J. & Shirley, L. W. (2003). Social vulnerability to environmental hazards. *Social Science Quarterly*, *84*(2), 242–261.

Davidson, T. & Davidson, J. R. (2009). Bell hooks, white supremacy and the academy. In M. Davidson & G. Yancy (Eds.), *Critical perspectives on bell hooks* (pp. 68–81). New York, NY: Routledge.

Drury, J., Cocking, C., & Reicher, S. (2009). The nature of collective resilience: Survivor reactions to the 2005 London bombings. *International Journal of Mass Emergencies and Disasters*, *27*(1), 66–95.

Enarson, E. (2007). Identifying and addressing social vulnerabilities. In W.L. Waugh & K. Tierney (Eds.), *Emergency management: Principles and practice for local government* (pp. 257–278). Washington, DC: ICMA Press.

Enarson, E., Fothergill, A., & Peek, L. (2007). Gender and disaster: Foundations and directions. In H. Rodríguez, E. Quarantelli, & R. R. Dynes (Eds.), *Handbook of disaster research* (pp. 130–146). New York, NY: Springer.

Fordham, M., Lovekamp, W. E., Thomas, D. S. K., & Phillips, B. D. (2013). Understanding social vulnerability. In D. S. K. Thomas, B. D. Phillips, W. E. Lovekamp, & A. Fothergill (Eds.), *Social vulnerability to disasters* (pp. 167–198). Boca Raton, FL: CRC Press.

Kenney, C. M. & Phibbs, S. (2015). A Māori love story: Community-led disaster management in response to the Ōtautahi (Christchurch) earthquakes as a framework for action. *International Journal of Disaster Risk Reduction*, *14*(1), 46–55.

Klinenberg, E. (2003). *Heat wave: A social autopsy of disaster in Chicago*. Chicago, IL: University of Chicago Press.

Lambert, S. (2014). Indigenous peoples and urban disaster: Māori responses to the 2010-12 Christchurch earthquakes. *Australian Journal of Disaster and Trauma Studies*, *18*(1), 39–48.

Lee, S. (2012). Experiences of earthquake and future discussions for multicultural families (translation mine). In H. Komai & E. Suzuki (Eds.), *The Great East Japan earthquake and immigrants* (translation mine) (pp. 56–74). Tokyo, Japan: Akashi Shoten.

Leong, K. J., Airriess, C., Chen, A. C., Keith, V., Li, W., Wang, Y., & Adams, K. (2007). From invisibility to hypervisibility: The complexity of race, survival, and resiliency for the Vietnamese-American community in Eastern New Orleans. In K. A. Bates & R. S. Swan (Eds.), *Through the eyes of Katrina: Social justice in the United States* (pp. 171–188). Durham, NC: Carolina Academic Press.

Lindell, M. K. (2013). Disaster studies. *Current Sociology Review*, *61*(5–6), 797–825.

Marlowe, J. & Lou, L. (2013). The Canterbury earthquakes and refugee communities. *Aotearoa New Zealand Social Work Review*, *25*(2), 58–68.

Matthewman, S. (2015). *Disasters, risks and revelation: Making sense of our times.* Hampshire, UK: Palgrave Macmillan.
McIntosh, P. (2007). White privilege: Unpacking the invisible knapsack. In M. L. Anderson & P. H. Collins (Eds.), *Race, class and gender: An Anthology* (6th ed., pp. 98–102). Belmont, CA: Thomson Wadsworth.
Morrow, B. H. (1999). Identifying and mapping community vulnerability. *Disasters, 23*(1), 1–18.
Morrow, B. H. (2008). Community resilience: A social justice perspective – CARRI Research Report 4. *Community and Regional Resilience Institute.* Retrieved from http://www.resilientus.org/wp-content/uploads/2013/03/FINAL_MORROW_9-25-08_1223482348.pdf
Peacock, W. G., Morrow, B. H., & Gladwin, H. (1997). *Hurricane Andrew: Ethnicity, gender and the sociology of disasters.* New York, NY: Routledge.
Peacock, W. G., Van Zandt, S., Zhang, Y., & Highfield, W. (2015). Inequalities in long-term housing recovery after disasters. *Journal of the American Planning Association, 80*(4), 356–371.
Phillips, B. D. (2014). *Qualitative disaster research: Understanding qualitative research.* New York, NY: Oxford University Press.
Rodin, J. & Garris, R. (2012). Reconsidering resilience for the 21st century. *USAID Frontiers in Development* (pp. 110–120). Retrieved from https://www.usaid.gov/sites/default/files/documents/1868/JudithRodinandRobertGarris.pdf
Santos-Hernández, J. M., & Morrow, B. H. (2013). Language and literacy. In D. S. K. Thomas, B. D. Phillips, W. E. Lovekamp, & A. Fothergill (Eds.), *Social vulnerability to disasters* (pp. 265–280). Boca Raton, FL: CRC Press.
Skutnabb-Kangas, T., & Phillipson, R. (1995). *Linguistic human rights: Overcoming linguistic discrimination.* Berlin, Germany: Mouton de Gruyter.
Solnit, R. (2009). *A paradise built in hell: The extraordinary communities that arise in disasters.* New York, NY: Viking.
Tierney, K. (2009). Disaster response: Research findings and their implications for resilience measures – CARRI research report 6. *Community and Regional Resilience Institute.* Retrieved from http://www.resilientus.org/wp-content/uploads/2013/06/Tierney-20092.pdf
Tierney, K. (2014). *The social roots of risk: Producing disasters, promoting resilience.* Stanford, CA: Stanford University Press.
The Great East Japan Disasters Korean Immigrants Interview Project (translation mine). (2015). *Experiencing disasters in foreign land: 3/11 for Koreans living in Tohoku* (translation mine). Miyagi, Japan; Ara Emishi.
Turner, V. (1969). *The ritual process: Structure and anti-structure.* Chicago, IL: Aldine Publishing.
Uekusa, S., & Matthewman, S. (2017). Vulnerable and resilient? Migrants and refugees in the 2010-2011 Canterbury and Tohoku disasters. *International Journal of Disaster Risk Reduction*, 22: 355–361.
Webb, G. R. (2007). The sociology of disaster. In C. D. Bryant, & D. L. Peck (Eds.), *21st century sociology* (pp. II276–II286). Thousand Oaks, CA: Sage Publications, Ltd.

CHAPTER 8

DISASTER EXCEPTIONALISM IN INDIA: THE VIEW FROM BELOW

José Manuel Mendes

ABSTRACT

The purpose of this chapter is to analyze disaster exceptionalism in India, focusing on the case of Kosi river floods in the State of Bihar and their impact on Dalit communities. Data were gathered through document analysis and a qualitative approach based on interviews with national and local leaders and activists of NGOs and Dalit organizations. The main finding is that there are no second-generation social movements related to disasters in India, mainly in what concerns Dalit discrimination. The Disaster Management Act of 2005 reinforced the centralized and top-down nature of the Indian state concerning disasters. On the other hand, national Dalit organizations like National Confederation of Dalit and Adivasi Organisations (NACDAOR) and National Dalit Watch do not possess the expertise to alter the approach to disasters from a contingent and exceptional one to a more structured and long-term perspective.

The chapter shows how extreme events and permanent hazardous situations tend to increase the legitimacy of state intervention, often involving the suspension of social and economic norms, creating a state of exception, which indicates the inevitable presence of the state. The abyssal line

Recovering from Catastrophic Disaster in Asia
Community, Environment and Disaster Risk Management, Volume 18, 145–160

ISSN: 2040-7262/doi:10.1108/S2040-726220160000018007

that separates those individuals and groups that are integrated from those defined as disposable and invisible crosses through both the Global South and the "little colonies" of the North, reinforcing the logic of states that want to be bigger and stronger than their own citizens.

Keywords: Disaster exceptionalism; state; India; Bihar; floods; Dalit

On its online edition of August 25, 2016, the news magazine *India Today* reported that Nitish Kumar, Chief Minister of India's Bihar state, visited flood-affected areas in Patna, the state capital, where thousands of people were displaced to relief camps because of Kosi river's floods (Rohit Singh, 2016). It was a reliving of the Kosi river's floods of August 2008, when more than three million people were affected and more than 200,000 houses were destroyed (Government of Bihar, World Bank and Global Facility for Disaster Reduction & Recovery, 2010).

In this chapter, I argue that extreme events, such as disasters and catastrophes, show the political work done to put disposable individuals and groups outside social networks and national imagined communities (Mendes, 2010). It analyzes disasters impact and recovery for Dalits[1] in India, taking as a case study the 2008 River Kosi floods in the state of Bihar. Contrary to the conclusions of Luft (2009) for New Orleans, and in line with the notion of disasterscape proposed by Kapur (2010), there is no second-generation social movement organizations related to Dalits and disasters in India, as the analytical and mobilization framework is still one of disaster exceptionalism.

DISASTERS AND THE PERFORMATIVITY OF THE STATE

Extreme events and permanent hazardous situations (where risk is a condition and not an event, Santos, 2009b), both in the Global North and South, tend to increase the legitimacy of state intervention, often involving the suspension of social and economic norms, creating a state of exception, which indicates the inevitable presence of the state.

My argument is that state is the mediator that legitimates the integration of societies into global capitalism, a capitalism that becomes more and more a disaster capitalism (Klein, 2007), not only owing to natural and technological disasters. The abyssal line (Santos, 2014, 2007) that separates those individuals and groups that are integrated from those defined as disposable and invisible crosses through both the Global South and the "little colonies" of the North, both through the regulation/emancipation logics present in the North as through the appropriation/violence logics that prevail in the Global South.

The advent of modernity led to the concept of *fortuna* being replaced by the concept of risk, expanding the context of trust, in addition to the role of the state in regulating this area (Giddens, 1991; Luhmann, 1993). However, as Santos has argued (1995), we are also witnessing a widening gap between the capacity to act and the capacity to foresee, thus dramatically increasing risks, both in terms of scale and frequency. This presents new challenges for regulation risk by states, due above all to transnational dynamics and the increasing needs of citizens in terms of security and the existence of well-defined prevention and mitigation plans with clear objectives (Hood, Rothstein, & Baldwin, 2001).

The theorization of "extremes" is based on an ecology of fear (Davis, 1998) and a politics and culture of fear (Furedi, 2005, 2002), which does not allow for the progressive and gradual composition of a shared world (Latour, 2005). Extreme events such as disasters and catastrophes show the political work involved in positioning disposable groups and individuals outside social networks and communities within national imaginaries.

The alternative is to define social technologies for participation leading to the construction of civic epistemologies that allow for the informed and critical presence of citizens within public space (Jasanoff, 2005). It is, as stated by Santos (2014, 2007), to confront the monoculture of science with the ecology of knowledge, that is, the plurality of heterogeneous knowledge.

Civic epistemologies define how democratic societies acquire shared knowledge for the purposes of collective action, shaped by different political cultures and national contexts.

The consolidation of the importance of civic epistemologies must be based on experience accumulated through analysis of disaster situations or catastrophes in which, contrary to the affirmations of many specialists and political leaders, the responses of the populations, groups, and individuals affected are not based on irrationality or panic (Clarke, 2005) but on mutual help, solidarity, and the construction of disaster communities that shape the capacity to resist, recover, and slowly reestablish bonds, lived experiences, and inhabited spaces.

The historical geography of disasters shows evidence of the hegemonic and performative production strategies of an insecure world. As Bankoff (2001, 2004) clearly demonstrates at the end of the 20th century, the *topos* of natural disasters replaced the *topoi* of tropicality (disease is resolved by Western medicine) and development (poverty is resolved by investment and western aid) to produce another dangerous and distant exoticism.

My thesis is that catastrophes and extreme events reveal the political nakedness of people, in which political powers separate political beings (citizens) from naked life (bodies). Catastrophes lead to exoticization processes, dispositives, and discourses, based on racial, class gender, or caste criteria, and to biopolitics and the geo-economic management of populations.

Therefore, and recurring to Agamben's (1998) concept of *homo sacer*, more than the inclusion of the excluded, *homo sacer* is the real condition of all of us. Both in North and South, not forgetting centuries of oppression and colonial domination, all citizens integrated in states and/or nations can find, in the diffractive game of images, discourses, and technologies related to extreme events and catastrophes, the weakness of the state as a global imaginary therapist of its citizens. In extreme events and catastrophes, we find states that want to be bigger and stronger than their own citizens.

DISASTERS IN INDIA: THREE ANALYTICAL APPROACHES

The majority of the literature on disasters in India focuses only on certain regions or states, and global and integrated approaches to the subject are lacking.[2] Nevertheless, we can discern three main currents: an economic historical analysis of disasters in a liberal framework, represented by T. Roy; a postcolonial *demarche* that, although based on historical analysis, accentuates the ideological and political continuities in disaster management, as can be seen on the work of R. D'Souza; and a more institutional and geographical proposal, based on a critical analysis of disaster prevention and mitigation from vulnerable populations' perspectives and the social and economic impacts on victims. A. Kapur mainly represents this latter approach.

For Roy (2012), and mainly concerning natural disasters as famines, storms, and earthquakes, in the long run and in a historical perspective, more than the profit motive or reasons of power for the colonial period, what stands out in India is the scientific enterprise, that is, knowledge for power over nature (2012, pp. 145–153). Although the author recognizes the impact of structural factors related to gender or caste (2012, p. 149), he clearly states that his definition of natural disaster is not based on a "victim" perspective (2012, pp. 12–14) and an emphasis is given to the theoretical understanding of natural disasters.

This perspective is in contrast with D'Souza's (2006, p. IX) approach that emphasizes the long-term political impact of a flood-dependent narrative to a flood-vulnerable landscape. This resulted in flood control as a narrative about political economy and not as a list of technical interventions. For D'Souza, we have to be attentive to "specific colonial momentum as a social form with regard to ecological change" (2006, p. 4). The challenge is to analyze how the colonial momentum works in a postcolonial context, mainly in what relates to minorities and vulnerable groups.

From a geographical and institutional perspective, Kapur emphasizes the long track in major disasters and catastrophes in India, in what she names as a "grim reality" (2005a). Kapur analyzes the disaster institutional configuration in India after independence and highlights that it had a fragmented nature and after the event orientation until the 1990s. The increase in number and impact of natural and technological disasters, as well as the pressure of globalization and international institutions, mainly the United Nations, resulted in a more integrated institutional framework.

Kapur (2005b) also argues that there is a distressing picture of the insensitivity of Indian society to disaster prevention and management, resulting in a careless approach to victims, even after the 2004 tsunami and the institutional and legal changes originated by the Disaster Management Act of 2005 (DMA-2005). The emphasis on natural agents as the causes of disaster led to a technocratic approach and on managerial disaster response when controls failed (Parasuraman, 2013, p. 7). Even DMA-2005, according to Parasuraman, did not incorporate sufficiently the notion of vulnerability and its interplay with hazards in its conceptualization of disaster (2013, p. 8). For Kapur, this demands a change in paradigm, assuming that nature is neutral and that socioeconomic, cultural, and political factors are related to disasters, as "the composition of deaths across different geophysical extremes reveals that the fatalities are extremely selective" (Kapur, 2005b, p. 4560).

To accentuate the specificities of disasters in India, Kapur (2010) coined the term *disasterscape*, following the use of the suffix "scape" and its consequent analyses proposed by Appadurai (1996). Disasterscape relates to the collective condition of disaster. Disasterscape is also a commodity, which cannot only be sold but also politicized and corrupted (Kapur, 2010, p. 5).

In what concerns disasters and citizenship in India, Chhotray (2014, p. 224), concludes that, more important than allocating specific rights to more disaster-affected individuals or communities, "any serious attempt to address the injustices of disasters cannot be restricted to making disaster relief a justiciable right, and must involve a larger conversation about the role of the state in improving the conditions of just social citizenship, whether through the public provision of welfare or the pursuit of redistributive policies. This is the biggest lesson of viewing disaster relief as an arena of justice."

DISASTERS AND DALITS

The specific cases in India reported in this chapter are related to the radical and paradigmatic change following the enactment in this country of the Disaster Management Act of 2005 (DMA-2005). This national legal initiative implemented a state structure for the management of disasters, involving national, regional, district, and local authorities (*panchayats*).

Although the National Institute for Disaster Management was created in 2003, the 2004 tsunami effects led to radical changes in India's approach to disasters. The 2004 tsunami was, in many ways, a founding event in India. It saw India refusing international relief aid and inaugurating also its ambition as a regional power in disaster management. At the same time, the tsunami induced a radical change in disaster management in India with the enactment of the DMA-2005. This reinforced the role of the Indian State in the definition of disaster prevention strategies and mitigation (Patil, 2012) and radically changed the institutional culture of disaster (Bankoff, 2006).

DMA-2015 reinforced the role of the Indian State in the definition of disaster prevention strategies and mitigation. That is, India entered in the rational and "civilized" world of risk management, after having refused international help in the aftermath of 2004 tsunami. India was now fully integrated in what Klein (2007) describes as disaster capitalism.

Inequalities in disaster support and recovery policies following the 2004 tsunami were documented and followed by the National Campaign on Dalit Human Rights.[3] Although many scholars in the field of subaltern studies have criticized the discourses of citizenship and human rights (Chatterjee, 2004), associating them with modernity and European enlightenment tradition (Chakrabarty, 2000), this Dalit organization conveys to this language in order to project internationally their grievances and claims.[4] Dalit claims and strategies can be analyzed as a counter-hegemonic human rights struggle, in line with the theoretical proposals of Santos (2009a).

The creation of the National Dalit Watch (NDW) in 2009, an initiative of the National Campaign on Dalit Human Rights, was based on the experiences of Dalit Watch-Bihar (DW-B), mainly during the 2007 and 2008 Kosi river floods. The narratives and discourses produced by the National Dalit Watch (2010) related to disasters must be read in an intercultural perspective that, as Santos (2009a, p. 25) proposes, "seeks both to strengthen the legitimacy of world-wide politics of human rights and to radicalize the struggles that can be undertaken in their name. The notion of interculturality is meant to convey the idea that the focus of intercultural exchanges is interpretation, the production and sharing of meaning."

India has a long history associated with catastrophic floods. The politics of flood control is deeply associated with its colonial heritage (D'Souza, 2006; Hill, 1997; Praveen Singh, 2008). More recent floods in Bihar have been denounced as resulting from top-down policies (Mishra, 2008). A grassroots NGO, Barh Mukti Andolan (BMA), has been founded by Dinesh Kumar Mishra. It proposed that, instead of relying on dams and embankments, floods management must resort to local leaders ("floods historians"), giving them tools for managing and coping with floods. In this sense, local communities should recapture the cultural and political control and ownership of rivers in India.

Lessons from the Field

My first contact with a Dalit organization was in New Delhi in October 2014 with the National Confederation of Dalit Adivasi Organisations (NACDAOR), the most representative entity in India. After some interviews with the leader and other activists and representatives, the conclusion was that disasters and their impact on Dalit communities was not a priority for the confederation. NACDAOR focused its activities mainly against atrocities, right to work and justice. In PT[5] analysis, one of the leaders of NACDAOR, the impact of disasters on Dalits is a direct consequence of geography and social discrimination due to caste, a fatalist discourse that does not engage direct action, social movements, or grievance actions related to Dalit disaster-affected communities. In his words

> The geography of villages, Dalits live in the most difficult part of the villages. Difficult to reach, and they will live around the area which is prone by disaster. They live nearby the drains, they live nearby the rivers, they live outside the villages, where there is a field, there is dirt, garbage and everything. So these places themselves are not accessible. So that also, you know, that also contributes to the thing. But most of it is the attitude. If I know you, and if I don't know a person and both of you are affected by the disaster, who will I save first?

The NDW, on the contrary, claims to be the only organization explicitly working on the inclusion of Dalits in disaster in India. It has developed an inclusive specific methodological tool for assessment and evaluation (Paul & Acharya, 2013). Since its inception, it has been present in major disasters in

India advocating for Dalits as the more affected communities and for equal and better disaster relief from the Indian National Government and the states.

Rajesh Singh, one of the program coordinators at NDW, summarizes their strategy in three spheres[6]: continuous evidence building using their specific methodological tool to evaluate whether exclusion is happening; using the collected evidence to mobilize local community leaders and to train them, and then start advocating on their entitlements on state and government regulations pertaining to disasters and training volunteers and Dalit organizations on disasters and exclusion; and, lastly, to try to connect to national, state, and district authorities to modify their current guidelines and texts and include caste based discrimination.

And Rajesh Singh explicitly argues that "the theory of no loss is dangerous for Dalit," for if there is no clear material or infrastructure damages or media amplification, less attention will be given by authorities to disaster impact on Dalit communities.

BIHAR 2008 FLOODS AND ABSENCE OF DISASTER SOCIAL MOVEMENTS

Bihar 2008 floods, one of the worst in India history, turned visible what is a normal accident (Perrow, 1999) for one million residents that live trapped inside Kosi river embankments and that every year can stand months of flooding effects due to waterlogging. Of these one million residents, 15% are Dalits (Mishra, 2008, p. 140). According to Dinesh Mishra, one of the leading experts on rivers in India, the 2008 events were caused by the eighth breach in the embankments since their construction in the 1950s.[7] He, nevertheless, highlights that, although Dalits are the most affected, other communities like farmers, laborers, and fishermen are deeply affected in their livelihoods and thousands have to migrate.[8] The media relevance and the intervention of international aid agencies in 2008 were due to the embankment breach having affected populations and communities of middle and high castes and social classes, as the River Kosi diverted its normal path by more than 160 km. This exceptionalism, attracting national and international relief, triggered an official response with the nomination by the Government of Bihar of a one-man Commission of Inquiry chaired by Justice Rajesh Balia, Retired Chief Justice, Patna High Court, to enquire into the causes of breach in the Eastern Kosi, so as to prevent occurrence of such disasters in the future.[9]

Civil society, through local activists and NGO leaders, also mobilized itself to document and report Dalit discrimination during 2008 Bihar floods. The invisibility of yearly flood affected Dalit communities, on the basis of the abyssal line of caste, due to the Kosi river's diversion was for that year put aside, although the abyssal line was maintained in what relates to disaster relief and reconstruction. With the support of the South African–based NGO, Action Aid, the local NGO, *Praxis*, and DW-B, a report was produced, on the basis of participative methodologies that documented caste-based discrimination in relief support and rebuilding, poor government relief coordination, and support (ActionAid, 2008).

As DW-B was the inspiration for NDW in formulating its objectives and participative methodologies, it was important to evaluate the continuity of its coalition of NGOs, organizations, and local movements to test the hypothesis of disaster exceptionalism in Bihar. For this purpose, I interviewed in 2014 three of the main actors of this coalition.

The overall conclusion is that the NGOs and movements coalition was disintegrated, each organization and leader continuing their Dalit advocacy and empowerment in separated and sparse activities. BMA, led by Dinesh Mishra and that congregated more than 700 local groups, had also dissolved and had no real expression in the field in 2014. Disaster exceptionalism, as what concerns Dalit communities, was the rule, with no articulation with NDW or NACDAOR. In disaster prevention, mitigation, and recovery in Bihar, the reality is a presence of a state stronger than its citizens.

Disaster and Human Rights

The first interviewee in Bihar was RS.[10] He is very knowledgeable and is a recognized expert on Ganges and Kosi rivers and their dams and embankments. For him the possibility of creating DW-B originated with DMA-2005. Although DMA-2005 reinforced the power of the Indian national and state governments, marginalizing NGOs, local authorities, and social movements, its definition of disaster allowed, according to RS, disasters and relief actions becoming a question of human rights, mainly for the international context and projection of activities, and at a national and state level, as citizen rights. Alongside the conventional narrative about Dalit atrocities, the Act allowed for mobilization on the basis of the grammar of human and citizen rights.

For RS, DW-B functioned as a pressure group, documenting caste inequalities caused by the Kosi floods thorough local surveys and claiming for grievances in a citizen rights framework to lower and higher courts, capitalizing on judicial activism in India (Pathe, 2001). In the interviewee's opinion, in Bihar and India the higher courts are more prone to civil rights demands and to judicial activism than the lower courts as these are more influenced by local interest groups and official policies.

Floods and disasters in general, according to RS, are seen by NACDAOR as having no political impact, relevance or projection, as it focuses mainly on atrocities and discrimination committed upon Dalits. NACDAOR does not have the technical expertise or awareness to discuss the structural problems related to floods, directly articulated to dams, engineering, embankments, and public policies concerning disaster prevention and management.

The Force of the Status Quo

Anindo Banerjee is the leader of *Praxis*, a Bihar local NGO specialized in participatory evaluation and assessment methodologies. His is a pragmatic and fatalist view, highlighting the power of the majority against the Dalit minority, of the established powers, based on heavy infrastructures and engineering and big public investments, against the demands and grievances of local communities stranded between Kosi's two embankments. Some communities and movements have organized public demonstrations against the construction of bridges and other infrastructures that increase the risk of flooding. However, there is in Bihar a huge political economy of the flood control industry. There was also some public interest litigations presented to lower and higher courts of Patna with no evident effect.

Furthermore, according to Banerjee, Bihar is a very centralized state, with only 7.5% of tax revenue reverting to local communities or *panchayats* and with almost no decision devolution to these communities. During the 2008 floods, the average lack time for any relief camp to be set up was 18 days. Before the setup of relief camps, people survived because they had their own mechanisms; they had their own safety nets and solidarity networks.

There was an attempt in 1987 in Bihar to counter disaster exceptionalism, when the state government created the Kosi Sufferers Development Authority. This Authority had as a mandate to relocate communities and assign land to those affected, assuring access to public jobs, and also rebuilding public buildings such as schools and other infrastructures. As the Authority congregated 19 government departments, coordination was a hindrance, and no real public policy or initiative was materialized in the communities living inside the embankments.

For Anindo Banerjee, the way to construct civic epistemologies on Kosi river floods and counter the official narratives is to resort to participatory methods. He exemplified with the project *Voices of the Kosi River*. A team of activists and experts travelled down the river, spending time with the local communities and using visual tools, participatory methods, and local mapping with location of resources, mobile resources, and habitations. This resulted in a mapping of local issues resourced and complemented with more traditional vulnerability analysis.

Women, Caste, and the Most Invisible Flood Victims

AB is recognized in India and internationally by her work with Dalit communities in Bihar. AB is keen to point Kosi 2007 floods, which mainly affected communities within the embankments and remained localized and reproduced discrimination based on the abyssal line of caste, as the event that alerted Dalit communities for how floods can be so damaging to their livelihoods, accentuating poverty and dispossession.[11] In 2008, DW-B and AB's organization took more than 100 volunteers to the field, and with the knowledge and knowhow of 2007, but with no material resources, started just helping people to find relatives and to mediate with local and state authorities. In AB's words:

> Now, when you look at the Dalits in the relief camps also, we realize that Dalits are at the last end. If you stand in line to get breakfast or the meals, they will be pushed and pulled out, and they will be at the last bit. And particularly women, I realized they would just gather together waiting for something to come. So, in many places I told them: "You can't stand at the back and expect that they are going to come to you. You have to push ahead with the rest of the crowd and get you what the solution you have to get." So motivating them, and I myself stood with them in the line and went ahead with them, so that, you know, they, so like that lot of and then they come across camps there, children are crying like anything because they are hungry in the morning.

AB helped many women to fill untouchability complaints, mainly pregnant women, to Bihar's State government. The objective was to obtain in the relief camps safe places so women could rest and get fed properly. For AB, as DW-B was a coalition of five organizations, totally based on voluntary work, it had no means to maintain its activities and build a structural, non-exceptional approach to disaster preparedness and mitigation. In the interview, she assumed a very critical position of NDW and its policy mainly based in New Delhi, and, in her words, the notion that "I felt that the emergency was over." Living in Patna for more than 20 years, AB has gained certain insights into their struggle and into their precarious life. For AB, only education can lift Dalits out of their condition and be the basis for their empowerment and for a structural approach to Kosi floods vulnerability.

CONCLUSION

In this chapter, I argued that disasters and extreme events are occasions where states are tested and activate political strategies to become stronger than their own citizens, putting disposable individuals and groups outside social networks and national imagined communities. This disaster exceptionalism can be countered by strong disaster-related social movements that mobilize alternative knowledge and practices in a structured and sustained way to advocate for more inclusion and equality.

Taking as a case study the river Kosi 2008 floods in the Indian state of Bihar, I showed how the coalition of movements and organizations congregated in DW-B dissolved itself, leaving local communities at the mercy of nature and stranded inside the river's embankments, in a fatalism of recurring floods. As disaster is seen as fatality by the major India Dalit confederation, NACDAOR, floods and other disasters reinforce the logic of disaster exceptionalism by India's national and state governments, in spite of changes in the national law and the implementation of national plans operating for disaster preparedness, mitigation, relief, and reconstruction. We can conclude the inexistence in Bihar, and India in general, of second-generation disaster social movements able to demand a structural approach to disasters and extreme events from the national and state-level authorities. This is accentuated due to the effect of the abyssal line that separates those individuals and groups that are integrated from those defined as disposable and invisible, known before as "untouchables" and now as "Dalit."

NOTES

1. The term Dalit is the political designation reclaimed by the representatives of the persons and communities that were known as Untouchables. The official designation consecrated in India's Constitution is Scheduled Castes.
2. For a recent overview of disaster risk management in India, see Garge, Ha and Khoo (2016).
3. For a general view of vulnerability and disasters in India, see Andharia (2013).

4. For the use and contradictions of a human rights framework in New Orleans Katrina first and second generation social movements, see Luft (2009).
5. To maintain interviewees anonymity when asked, and pseudonyms being difficult to assign in the context of India, where a name denotes caste, region, and even social position, I opted for this solution of attributing letters to identify them. PT. Personal interview with the author. New Delhi, October 12, 2014.
6. Rajesh Singh. Personal interview with the author. New Delhi, October 29, 2014.
7. Dinesh Mishra. Personal interview with author. Patna, November 28, 2014.
8. For the argument that disaster caused by Kosi floods is not an accidental interruption, but is linked to the socio-political structure of Bihar state, see Jha (2015).
9. The documents and reports of the Commission can be found on the electronic site http://www.kosi-aayog.bih.nic.in/. The Government of Bihar also produced an official report on the floods and needs assessment (Government of Bihar et al., 2010).
10. RS. Interview by the author. Patna, November 23, 2014.
11. AB. Personal interview with author. December 4, 2014.

ACKNOWLEDGMENTS

This chapter has been written as part of the research project ALICE—Strange Mirrors, Unexpected Lessons, financed by the European Research Council, and coordinated by Boaventura de Sousa Santos at the Centre for Social Studies, University of Coimbra, Portugal.

REFERENCES

Agamben, G. (1998). *Homo Sacer. Sovereign power and bare life*. Stanford, CA: Stanford University Press.

ActionAid (2008). *Bihar floods 2008. Needs assessment report*. Bangalore: ActionAid. Retrieved from https://www.globalgiving.org/pfil/2267/projdoc.pdf

Andharia, J. (2013). Vulnerability and disasters. Conceptual contours of a people-centred view. In S. Parasuraman & U. Krishnan (Eds.), *India. Disasters report II. Redefining disasters* (pp. 21–34). New Delhi: Oxford University Press.

Appadurai, A. (1996). *Modernity at large: Cultural dimensions of globalization*. Minneapolis, MN: University of Minnesota Press.

Bankoff, G. (2001). Rendering the world unsafe: 'Vulnerability' as Western discourse, *Disasters*, *25*(1), 19–35.

Bankoff, G. (2004). The historical geography of disaster: 'Vulnerabilility' and 'local knowledge' in Western discourse. In G. Bankoff, G. Farks, & D. Hilhorst (Eds.), *Mapping vulnerability. Disasters, development and people* (pp. 25–36). London: Earthscan.

Bankoff, G. (2006). *The culture of disasters. Society and natural hazard in the Philippines*. London: Routledge.

Chakrabarty, D. (2000). *Provincializing Europe. Post-colonial thought and historical difference*. Princeton, NJ: Princeton University Press.
Chatterjee, P. (2004). *The politics of the governed: Reflections on popular politics in most of the world*. New Delhi: Permanent Black.
Chhotray, V. (2014). Disaster relief and the Indian State: Lessons for just citizenship. *Geoforum, 54*, 217–225. doi:10.1016/j.geoforum.2014.01.013
Clarke, L. (2005). *Worst cases. Terror and catastrophe in the popular imagination*. Chicago, IL: University of Chicago Press.
Davis, M. (1998). *Ecology of fear. Los Angeles and the imagination of disaster*. New York, NY: Metropolitan Books.
D'Souza, R. (2006). *Drowned and damned. Colonial capitalism and flood control in Eastern India*. New Delhi: Oxford University Press.
Furedi, F. (2002). *Culture of fear. Risk-taking and the morality of low expectation*. New York, NY: Continuum.
Furedi, F. (2005). *Politics of fear. Beyond left and right*. New York, NY: Continuum.
Garge, R. N., Ha, H., & Khoo, S. (2016). Disaster risk management and the role of the armed forces: Critical analysis of reactive disaster management in India. In H. Ha, R. Fernando, & A. Mahmood (Eds.), *Strategic disaster risk management in Asia* (pp. 49–67). New Delhi: Springer.
Giddens, A. (1991). *The consequences of modernity*. Oxford: Polity Press.
Government of Bihar, World Bank & Global Facility for Disaster Reduction & Recovery (2010). *Bihar Kosi flood (2008) Needs assessment report*. Patna: Government of Bihar. Retrieved from https://www.gfdrr.org/sites/gfdrr/files/publication/GFDRR_India_PDNA_2010_EN.pdf
Hill, C. (1997). *Rivers of sorrow: Environment and social control in Riparian North Bihar, 1770–1994*. Ann Arbor, MI: Association for Asian Studies.
Hood, C., Rothstein, H., & Baldwin, R. (2001). *The government of risk: Understanding risk regulation regimes*. Oxford: Oxford University Press.
Jasanoff, S. (2005). *Designs on nature. Science and democracy in Europe and the United States*. Princeton: Princeton University Press.
Jha, M. K. (2015). Liquid disaster and frigid response: Disaster and social exclusion. *International Social Work, 58*(5), 704–716. doi:10.1177/0020872815589388
Kapur, A. (2005a). *Disasters in India: Studies of grim reality*. Jaipur: Rawat Publications.
Kapur, A. (2005b). Insensitive India: Attitudes towards disaster prevention and management. *Economic & Political Weekly, 40*(42), 4551–4560.
Kapur, A. (2010). *Vulnerable India. A geographical study of disasters*. Shimla/New Delhi: Indian Institute of Advanced Study/Sage.
Klein, N. (2007). *The shock doctrine. The rise of disaster capitalism*. London: Allen Lane.
Latour, B. (2005). *Reassembling the social. An introduction to actor-network theory*. Oxford: Oxford University Press.
Luhmann, N. (1993). *Risk: A sociological theory*. Berlin: Walter de Gruyter.
Luft, R. (2009). Beyond disaster exceptionalism: Social movement developments in New Orleans after Hurricane Katrina. *American Quarterly, 61*(3), 499–527.
Mendes, J. M. (2010). Pessoas sem voz, redes indizíveis e grupos descartáveis: os limites da Teoria do Actor-Rede. [Voiceless People, Unnamable Networks, and Disposable Groups: the Limits of Actor-Network Theory], *Análise Social, XLV*(196), 447–465.

Mishra, D. K. (2008). *Trapped! Between the devil and deep waters*. Dehra Dun and New Delhi: Peoples' Science Institute (PSI) and South Asia Network on Dams, Rivers and People (SANDRP).
National Dalit Watch. (2010). *The uncertainties of life. Living through waters of dejection*. New Delhi. Retrieved from http://www.ncdhr.org.in/latestinterventions/Yamuna%20flood%20report.pdf.
Parasuraman, S. (2013). Rethinking disasters. Searching for clarity and relevance in disaster discourse. In S. Parasuraman & U. Krishnan (Eds.), *India. Disasters report II. Redefining disasters* (pp. 3–17). New Delhi: Oxford University Press.
Pathe, S. P. (2001). Judicial activism: The Indian experience. *Washington University Journal of Law & Policy*, *6*(29), 29–107. Retrieved from http://openscholarship.wustl.edu/law_journal_law_policy/vol6/iss1/3
Patil, P. (2012). Disaster management in India. *Indian Streams Research Journal*, *2*(1), 1–4.
Paul, L. M., & Acharya, B. (2013). *Inclusive vulnerability mapping & monitoring of post disaster response (IVM & MPDR). Towards developing a methodology and template for field practice (A process to ensure Dalit and gender inclusion)*. New Delhi: SWADHIKAR/National Dalit Watch - NCDHR.
Perrow, C. (1999). *Normal accidents: Living with high risk technologies*. Princeton, NJ: Princeton University Press.
Roy, T. (2012). *Natural disasters and Indian history*. New Delhi: Oxford University Press.
Santos, B. S. (1995). *Towards a new common sense. Law, science and politics in the paradigmatic transition*. New York, NY: Routledge.
Santos, B. S. (2007). Beyond the abyssal line: From global lines to ecologies of knowledges. *Review*, *XXX*(1), 45–89.
Santos, B. S. (2009a). If God were a human rights activist: Human rights and the challenge of political theologies is humanity enough? The secular theology of human rights. *Law, Social Justice & Global Development Journal (LGD)*. Retrieved from http://www.go.warwick.ac.uk/elj/lgd/2009_1/santos
Santos, B. S. (2009b). *Toward a post-colonial risk theory*. Conference at the Meeting *Risk and The State in a Globalized World*. Coimbra: Centre for Social Studies, 18 and 19 September.
Santos, B. S. (2014). *Epistemologies of the South. Justice against epistimicide*. Boulder, CO/London: Paradigm Publishers.
Singh, P. [Praveen] (2008). The colonial state, zamindars and the politics of flood control in Northern Bihar (1850–1945). *The Indian Economic and Social History Review*, *45*(2), 239–259.
Singh, R. [Rohit] (2016). Nitish visits flood-affected areas in Patna, takes stock of relief camps. *India Today*. Online edition, 25 August. Retrieved from http://indiatoday.intoday.in/story/nitish-kumar-flood-affected-areas-patna/1/748644.html

CHAPTER 9

THE 2015 NEPAL EARTHQUAKE: FROM RESCUE TO RECONSTRUCTION

Chandra Lal Pandey

ABSTRACT

Nepal is confronted by an increasing number of natural- and human-induced disasters, making it the most disaster-prone country in the world. Landslides, floods, droughts, fires, thunderbolts, blockades, and earthquakes, among others, occur frequently. All these disasters take a high toll on people and cause heavy damage to physical properties worth billions of dollars every year. One of the recent examples in the history of disasters was the devastating earthquakes that struck Nepal in 2015 taking away human lives, destroying physical infrastructures, altering cultures, challenging institutions, and devastating the hope and ambitions of people. Thirty-three of the Nepal's 75 districts were affected, at different degrees, and 14 out of 33 were badly damaged by the two major earthquakes on the 26th of April and the 12th of May 2015. This chapter provides a reflective dialogue on the current state of the disaster-recovery process with a focus on the delivery of recovery services aiming to (1) understand how key

Recovering from Catastrophic Disaster in Asia
Community, Environment and Disaster Risk Management, Volume 18, 161–176

ISSN: 2040-7262/doi:10.1108/S2040-726220160000018008

actors are responding to disaster-recovery works in Nepal and (2) identify knowledge gaps in the disaster response policy and research arenas in managing future disasters in Nepal.

Keywords: Disaster; Nepal earthquakes 2015; rescue; recovery; reconstruction

INTRODUCTION

Nepal is a Himalayan country located between India and China, covering an area of 147,181 sq. km. The country spreads 145 km north to south and 885 km from east to west. Nepal is divided into three geographical regions: the Himalayan, hill, and Terai (plain region). Terai covers 17%, the hill covers 68%, and the Himalayan region covers 15% of the total land area of Nepal. The Terai is home to about 49% of the population and is known as the breadbasket of Nepal. The total population of Nepal is approximately 29 million. Nepal ranks 11th globally as the most vulnerable country to earthquakes and is one of the most natural disaster-prone countries in the world (CDRMP, 2011; NSDRM, 2008; Paudel and Panthi, 2010). Landslides, floods, droughts, fires, thunderbolts, blockades, and earthquakes, among others, occur frequently. All these disasters take a high toll on people and cause heavy damage to physical properties worth billions of dollars every year.

As Nepal lies on the fault line of the Eurasian and Indian tectonic plates, the earthquakes and land tremor-induced disasters are frequent. The earthquake of 1255 AD, which was recorded for the first time, struck Nepal killing one-third of the population in Kathmandu Valley (GoN-NPC, 2015). One of the latest examples in the history of disasters was the devastating earthquakes that struck Nepal in 2015 taking away human lives, destroying physical infrastructures, altering cultures, challenging institutions, and devastating the hope and ambitions of people. Thirty-three of Nepal's 75 districts were affected, at different degrees, and 14 out of 33 were badly damaged by the two major earthquakes on the 26th of April and the 12th of May 2015. These earthquakes, as well as hundreds of aftershocks that followed, killed nearly 9,000 people, injured more than 22,000, and destroyed or damaged over half a million houses, as well as many historic monuments and public infrastructure (GoN, 2015). Post-disaster needs assessment (PDNA) estimated that the total value of disaster effects (damages and losses) caused by the 2015 earthquakes was about US$7 billion, which is approximately equivalent to one third of Nepal's gross domestic product (GoN-NPC, 2015). With a highly differentiated demographic composition of society, disasters in Nepal have particularly affected the poor, women, and marginalized groups. The impact of the mega earthquakes was severe because Nepal was ill-prepared for a calamity of such scale and because it struck the country at a time of drafting a new constitution after the prolonged political instability following the decade long civil war.

Following the first earthquake of April 2015, responses came from the local communities, government, NGOs, business, and donors, but the efficiency and effectiveness of responses from different groups and organizations varied. Immediate response from the government was appreciated but the big donors were lost in the process and were almost unhelpful. Neighbors and communities were effective in the immediate search, rescue, and relief stages. However, serious questions are raised about the government capacity of raising and mobilizing resources effectively for reconstruction and the accountability of recovery and reconstruction leadership.

Over the past 18 months, the Government of Nepal (GoN) and a wide range of disaster-recovery stakeholders have undertaken a range of measures to respond to the disaster including immediate response to victims; relief and rehabilitation; cremation cost and assistance to the family of victims; temporary shelters; and reconstruction measures, plans, and activities. To expedite the reconstruction process, the GoN constitutionally formulated a special body known as National Reconstruction Authority (NRA) with a tenure provision of 5 year to develop sustainable infrastructure, to address the concerns and practical disaster needs of the quakes' victims, and to complete the reconstruction work damaged by the 2015 earthquakes (NRA, 2015).

Under the leadership of the Prime Minister, the NRA prepared a framework on post-earthquake reconstruction in line with the PDNA report aimed at completing reconstruction works in 5 years time. As advised by the GoN, the NRA decided to provide a package of NPR 200,000 in three installments to the quake victims for completion of reconstruction. In the first installment, the NRA decided to provide each of the victims a grant of NPR 50,000. For this, the victim needed to have the house map designed as per the set criteria in the building code and get approval. After foundation laying, the eligible victim would get NPR 70,000 each. They would get NPR 80,000 each in the last phase, for roofing the house. However, on September 9, while addressing the parliament meeting, Prime Minister Puspa Kamal Dahal announced an increase of another NPR 100,000 in the grant assistance amount to be provided to earthquake victims for reconstruction (Republica, September 9, 2016). Out of the grant amount of NPR 300,000, the NRA is providing NPR 50,000 to the victims in the first installment, NPR 150,000 in the second installment, and NPR 100,000 in the last (*The Rising Nepal*, October 5, 2016). The NRA seems to be working actively to achieve its mission, yet the voices and grievances of affected people are echoing ubiquitously.

Till the date of this writing, 20% of the earthquake victims from 11 quake-affected districts outside of Kathmandu have not received the first installment of NPR 50,000. Over 100,000 households from the 11 districts lodged complaints, claiming they were excluded from the list of beneficiaries for not having land certificates even though they were eligible to receive grant from the government (*The Himalayan Times,* August 16, 2016). The primary reason is that the victims do not have land ownership certificates, and there has not been a local elected body of governance since 2002 to address this problem effectively, efficiently, and legitimately. Not a single family has received the second installment of NPR 150,000 and the third installment of NPR 100,000 yet. The quake victims of the Kathmandu Valley are still on the waiting list. The prolonged delays in the distribution of reconstruction grants to the victims of the earthquakes solicit important questions about the causes and consequences of the delays to identify operational, institutional, and policy gaps for present and future reconstruction operations.

The main objective of this chapter is to provide a reflective dialogue on the current state of the disaster-recovery process with a focus on the delivery of recovery services and gaps. This chapter aims to (1) understand how key actors are responding to disaster-recovery works in Nepal and (2) identify operational, institutional, and policy challenges in disaster response and research arenas to conduct cutting-edge critical research for meeting the gaps in managing future disasters in Nepal.

UNDERSTANDING KEY ACTORS' RESPONSES TO DISASTER RECOVERY

Recurrent natural disasters pose a formidable challenge to human society's survival and prosperity. Developing countries are particularly vulnerable to natural catastrophes because of their low resilience capacity, which is exacerbated by poor governance and institutional capacity. Nepal is also exposed to multiple natural disasters and hazards (GoN, 2015). On Saturday, April 25, 2015, at 11:56 local time, a 7.8 magnitude earthquake as recorded by Nepal's National Seismological Centre (NSC), struck Barpak of Gorkha District, about 76 km northwest of Kathmandu. This was the most devastating earthquake in Nepal since the 1934 quake. According to the post-earthquakes PDNA 2015 estimates, the earthquake has pushed back more than 700,000 people below the poverty line (<1.25 US$/day) (GoN-NPC, 2015). Nepal is in need of investing substantial resources to "build back better," and the NRA estimates that US$9.3 billion is required for completing the reconstruction works. Although the full budget to address the disastrous damages of the 2015 earthquakes is still lacking, and the responses being delivered by the key actors including government, international and national nongovernmental organizations (I/NGOs), and donors working in the areas of disaster response are not satisfactory, the following tasks have been completed and more attempts are being made.

Search and Rescue Initiative

As mandated by the Natural Calamity Relief Act 1982, the Central Natural Disaster Relief Committee (CNDRC) recommended the GoN to declare a state of emergency to highly affected districts. The Cabinet declared emergency to 14 districts: Gorkha, Sindhupalchowk, Dhading, Kavre, Dolakha, Nuwakot, Ramechhap, Sindhuli, Rasuwa, Kathmandu, Lalitpur, Bhaktapur, Makwanpur, and Okhaldhunga.

The National Emergency Operation Centre (NEOC) was activated under the leadership of Ministry of Home Affairs (MoHA) to lead a central coordination point in response to disasters, coordinate between government and national and international organizations for relief support. MoHA appealed to national and international organizations, donors, and local communities to contribute in Emergency Operations Centres initiated at the national, regional, district, and municipality levels. A Response Coordination Center was also established to accelerate the Search and Rescue (SAR) operation. There were a total of 4,521 SAR personnel from 34 countries (Algeria, Australia, Belgium, Bangladesh, Canada, China, France, German, Hungary, India, Indonesia, Israel, Japan, Jordon, Malaysia, Mexico, Netherland, Norway, Oman, Philippines, Poland, Pakistan, Russia, South Korea, Sri Lanka, Singapore, Spain, Sweden, Switzerland, Thailand, Turkey, UAE, UK, and USA) coordinating with the National Security Forces (Nepal Army, Nepal Police, Armed Police Force) to rescue lives and recover dead bodies. The Multi-National Military Coordination Centre (MNMCC) under the leadership of the Nepal Army operationalized all foreign military assets through daily meetings in the Army Base, while the On-Site Operations Coordination Centre (OSOCC) was established by United Nations Office for the Coordination of Humanitarian Affairs (UNOCHA) at the base of operations, which was located at Tribhuvan International Airport.

On April 27, 2015, the cluster system was activated and monitored by the specific cluster leads and coleads. During the search and rescue period, the GoN effectively mobilized its administrative and security machineries including the Nepal Army, Nepal Police, Nepal Armed Police Force, and bureaucracy in a short span of time. The international community lent its support immediately. Besides, community support was crucial to tackle the situation. The GoN declared 14 districts as the crisis-hit districts for the purpose of prioritizing the rescue, relief, and rehabilitation process. The SAR tasks carried out by Nepal's security personnel were satisfactory, but commentators and public perceptions of the affected people demonstrated that the international response was centered in easily accessible areas and process was slow and inadequate as they failed to reach out in due time in remote, rural, and hilly areas (GoN-NPC, 2015, p. 20). Nepali people particularly the youths were also actively engaged in the search and rescue initiatives utilizing their indigenous knowledge, although they did not receive any standard training. Social media particularly Facebook and Google Crisis Response played an important role in sharing information about missing and safe persons as well as in terms of providing relief and helping raise awareness and funds. According to quake-survivors, the role of national security agencies, neighbors, and communities were found to be very important in the SAR initiative.[1]

Relief Distribution Initiative

As the search and rescue efforts were over, the focus shifted on the distribution of relief materials to bring earthquake-affected peoples' lives to normalcy. The key actors involved in the relief distribution initiative were government agencies, security personnel, people-to-people support, non resident Nepalis, and I/NGOs. The key items on the list were distribution of clean drinking water, food supplies, tents, medical care, and other emergency supplies to cities and remote areas. Connecting survivors to important information about reuniting individuals with their families, finding medical care for people in need, connecting people with agencies distributing emergency supplies, arranging safety inspections for damaged buildings, and getting children back to schools were some of the important tasks during this phase.

With the vast geographical area affected by the earthquake and distinct needs in each of the 14 severely hit districts, the earthquake relief distribution had to take place in accessible cities like Kathmandu valley and the largely inaccessible remote geographical environment from Kathmandu to the remotest and worst affected places of the hilly districts such as Gorkha, Dolakha, and Sindhupalchowk. The complex geographical terrain of Nepal and the poor infrastructure hindered the relief distribution efforts. The monsoon season from June to September in 2015 significantly downscaled air and land transport that constrained access to remote villages, and the fuel shortage caused by Indian blockade further delayed relief operations (UNOCHA, 2015). Given that an incomparably higher number of families in the rural areas were affected by the quakes than in the urban areas, the government or the concerned authorities had to prioritize the rural areas for relief distribution. Every family had to get relief materials such as tents, mats, and blankets for temporary shelter, clothes, utensils, and minimum amount of food for avoiding their existential crisis. Despite such dire needs, it was worrisome that only a small percentage of the affected people received relief materials on time, and most of the victims had to build their own *katero* (temporary shelters) at their own expense (*The Rising Nepal*, http://therisingnepal.org.np/news/3248). The quake victims complained for several months after the earthquakes that they were not provided with basic post-disaster requirements. They shared that most international NGOs and national NGOs only focused on the easily accessible areas. Even villagers in locations accessible besides roads only 30 miles from Kathmandu told that they were not contacted for immediate relief even after 1 month of the earthquakes (*The Guardian-UK*, May 3, 2015).

The GoN also made a decision to provide a cash winter relief package of Rupees 10,000 each to quake victim families in 2015; however, the process of relief distribution was still underway until February 2016, which means many of the victims did not get any support during the freezing winter season (*The Himalayan Times*, January 4, 2016). The media also highlighted about fake quake victims. The media reported that many people had their houses in cities as well as in the earthquakes devastated villages. According to government policy, those families who have two houses and one remains unaffected by the earthquakes are ineligible to claim relief and cash supports, but many of those households included their names in the list of earthquake-affected people through political gerrymandering and pocketed the cash. Even some local political leaders who did not suffer any loss in the catastrophic quakes received cash handouts, which were meant for victims, by using their political clout and producing documents identifying them as victims (*Republica*, July 7, 2015).

Reconstruction: Build Back Better

The 2015 earthquakes affected 33 out of 75 districts in Nepal. The GoN formulated the NRA in 2015 with the objectives of achieving on priority point four of the Sendai Framework—"Enhancing disaster preparedness for effective response and to 'Build Back Better[2]' in recovery, rehabilitation, and reconstruction" (United Nations, 2015, p. 14). The NRA has a tenure of 5 years to complete the reconstruction process based on the fundamental principle of "build back better." It has been over 18 months, and it appears that the actual reconstruction has just been initiated. The prolonged delay in the reconstruction process has trapped over 700,000 households in misery, awaiting a promised subsidy to rebuild their homes, consigning victim families to a second monsoon season and a second winter in temporary leaky shelters (*The New York Times,* April 30, 2016). According to an NRA official, the NRA completed the survey in 14 districts by mid-December 2016 (Thapaliya, 2016). After the earthquakes, the government distributed red cards and yellow cards to the victims, but the local media and the political parties raised the issue of the risk about card holding and management. Therefore, it was thought that PDNA had to be conducted to meet the international standards of relief distribution and reconstruction process. According to the NRA official, they conducted a household survey for PDNA in 11 districts where the census showed that 713,000 houses were affected but in three districts of Kathmandu valley they changed the modality and the verification model as the people were eagerly waiting for the grants (Thapaliya, 2016).

The NRA developed its own modality of grant distribution for reconstruction as follows:

1. Those victims whose houses were collapsed by the quakes and are willing to build a new house are eligible to receive NRS 300,000 in three transactions.
2. If the beneficiaries are in need of major fitting like retrofitting, they will be provided with NRS 100,000 in two transactions.
3. If the victims/house owners have already built their houses without waiting for the government grants and if they are in the list of beneficiary they will be given NRS 300,000 in one transaction.

According to NRA, the recovery and reconstruction process needs an amount of US$9.3 but they have US$4.1 billion as pledge and only US$2.78 billion has actually been received in the reconstruction coffer (Thapaliya, 2016). Nearly all of the pledged funds remain in the hands of the donors and unused (*The New York Times*, April 30, 2016). The government also developed a mechanism of one-door policy to monitor donors and I/NGOs activities and avoid duplication. However, the donors in the name of rescue and recovery had already spent US$66 billion while the work of reconstruction remains at large.

With regard to the progress of the reconstruction process, it was found that the housing grants have been distributed to 430,000 plus households. The government has constructed about 8,000 temporary schools, the development partners have built about 4,000 temporary schools, and about 10,000 permanent schools have been built so far. In the health sector, 200 prefab blocks have been completed, 168 are under construction and about 650 units are under preparation for construction. In the heritage sector, all 700-heritage structures are under preparation for construction and 20 public building are in the process of rebuilding. However, NRA in its reconstruction task encounters challenges of handling victims' grievances and financing. The NRA received 100,000 plus grievances of which 83,955 households were analyzed and only 57,685 household received grants. In the guidelines, it is stated that if the victims had one house in the name of their family members they are not entitled to get grants due to which nobody was satisfied with the survey and its outcome. Therefore, those who were not eligible to get into the record can ask for the reasons and if they are not happy with the reason or decision they can again give an application for reconsideration. With regard to financing, there is a gap between funds and needs.

In a half-day international workshop entitled "18 Months after the Nepal's Earthquakes: Practical Disaster Justice in the Recovery Work" convened by this author on December 6, 2016, at the Institute of Engineering (IOE), co-organized by the South Asia Institute of Advanced Studies, Nepal; Institute of Engineering, Nepal; Institute of Social and Environmental Transition, Nepal; and University of New South Wales, Australia, one of the victim representatives from Sidhupalchowk shared that "Rebuilding work has initiated with only few schools and public offices in the district and not a single private individual home has been even started to rebuild yet."[3] He further noted, "the fundamental problems of reconstruction are symmetrical building codes for urban and rural areas with no technical and financial support. The grant amount is so small that it cannot be enough to build a mud house while the government imposes expensive building codes for rebuilding houses." In Shankhu[4], many interviewees shared that many of the victims in the area do not have land certificate of the land on which they had built their houses in the past, and now due to lack of land certificate, they are ineligible to access the grants and soft loans. Many victims shared that they had already spent 20 months in tents, and it may take another 20 years for them to build the house in the current chaotic political realities and messy disaster politics.

The scale of the challenge Nepal facing to "build back better" is massive as there is a considerable gap between the amount of fund required and what is pledged, promised, and deposited in the government's coffer for reconstruction. Without clearly identifying the sources of funding, the reconstruction process remains a challenge as Nepal has a long track record of struggling to meet government expenditure targets even without any financial constraints imposed by recurring disasters. Besides, lack of institutional capacity and chronic corruptive practices in public institutions regularly plague projects and derail their effective completion. Furthermore, Nepal is trapped in prolonged political transition and there still remains significant political uncertainty in terms of federalizing the country and implementation of federal modality challenging the reconstruction concept of "build back better."

DISASTER MANAGEMENT INSTITUTIONS AND CHALLENGES

There were repeated alerts and predictions of an earthquake in Nepal primarily focusing in Kathmandu valley and its surrounding districts (Paudel, 2014). To mitigate or reduce risks and vulnerabilities of natural disasters including earthquakes, a number of initiatives, acts, and policies have been formulated in Nepal. The major existing disaster management institutions in Nepal are Natural Calamity (Relief) Act, 1982; Local Self Governance Act, 1999; National Strategy for Disaster Risk Management, 2009; Disaster Preparedness and Response Plan Guideline, 2011; National Disaster Response Framework, 2012; Local Disaster Risk Management Guideline, 2012; National Strategic Action Plan on Search and Rescue, 2014; and the Crisis Management Act of 2015 is under formulation. Also, there are four different fund mechanisms, which are the Prime Ministerial Disaster Relief Fund; Central Natural Disaster Relief Fund; Line Agencies Disaster Relief Funds; and Local Government Disaster Relief Funds for collecting national and international donation, aid, and grant and channelizing the fund to victims of disasters. Along with these initiatives, there have been a number of earthquake-related alerts, safety training, and demonstrations on a regular basis.

Yet, there was not a coherent legal and policy framework for disaster management in Nepal before "The Natural Calamity Relief Act in 1982." Prior to the passage of the Act, the relief and rescue work was carried out on an ad hoc basis. The Act was considered as a milestone for disaster management in Nepal as it outlines the establishment of various bodies, their roles, and responsibilities for disaster response at central, regional, and local levels; however, it is widely considered that this 34-year law needs to be updated to better meet current realities. The Local Governance Act 1999 clarifies the roles and responsibilities of the local government for responding to the development programmes or disaster situations. The legal documents are supported by policy frameworks such as the National Strategy for Disaster Risk Management and the National Disaster Response Framework, which were developed to provide more detail on the frameworks and systems across the whole disaster risk management spectrum—preparedness, risk reduction, response, and recovery and support a move toward a more holistic approach that is in line with commitments undertaken in the Hyogo Framework for Action 2005 and Sendai Framework for Action 2015.

Although there are a few disaster management institutions in Nepal, problems of cooperation, coordination, and mutual understanding between agencies have become serious challenges for implementation of existing legislations, which, however, still remain inadequate to cover the whole range of crosscutting issues of disaster management. Consequently, earthquakes victims did not receive timely effective rescue and relief services, with considerable delays and at times duplication. Connected to development, implementation of Nepal's national building code remains poor despite its formulation in 1994. Municipalities, responsible for issuing building permits, do not ensure compliance with the national building code. The de facto process of obtaining a building permit is very superficial and subjective with minimal compliance in practice. Few generalized checklists are developed and the questionnaires in the checklists are overlooked and are easily manipulated. Commentators comment that there is no effective mechanism for field verification of approved plans (Giri, 2013).

The post-earthquake disaster management policy goals of the GoN are to "build back better" and to develop capacity for resilience and building disaster resistant infrastructure. Building resilience is the capability of a social–ecological system to withstand or absorb disturbances including disasters without collapsing into a qualitatively different state and adds the capacity of humans to anticipate change and preparedness for future pathways (Gunderson & Holling, 2002; Holling, 1973; Walker et al., 2004). Resilience, for social–ecological systems, is related to (1) the magnitude of shock that the system can absorb and remain within a given state, (2) the degree to which the system is capable of self-organization, and (3) the degree to which the system can build capacity for learning and adaptation (Folke et al., 2002). However, it is argued that resilience's roots in engineering and the physical sciences has led to a lack of consideration of social aspects such as entitlements, power, access, equity, and scale, which are understudied elements (Weichselgartner & Kelman, 2014).

Reconstruction and resettlement processes have also been difficult in Nepal because of the divided society and partisan interests of political parties. Local politics and long-standing problems with the implementation of capital-spending projects are likely to prevent rapid rehabilitation of the affected areas. Longer-term policy challenges, including the setting-up of provincial and local governments and the provision of finance through cooperatives remain inadequately addressed and will hamper reconstruction efforts. Lack of clearly articulated roles, responsibilities, and limitations of local NGOs and international NGOs and lack of transparency in handling publicly accessible funds, i.e., the income and expenditure of the earmarked budget for assigned projects, have raised some prominent questions. Disaster victims, communities, and intellectuals question whether these organizations are delivering what they promise or the earmarked money is spent on their own purposes. Hence, Nepal's vulnerability to natural hazards can be seen as primarily a governance problem of local NGOs, international NGOs, and national public institutions (Piper, 2013). Future research needs to be directed toward understanding the challenges of institutional governance, institutional transparency, and accountability of the institutions working in the field of disaster management.

CONCLUSION

Despite significant investment in identifying and addressing vulnerabilities, there have been significant constraints. Among the 254 international NGOs registered in Social Welfare Council and working in Nepal, 61 highlighted that their focus was on disaster preparedness, disaster risk reduction, community development, and reconstruction among other issues (SWC, 2015). The GoN also allocates a large amount of budget in the disaster sector for disaster awareness and preparedness directly or indirectly. However, the objectives of disaster preparedness and disaster risk reduction have not been achieved at satisfactory standard. Instead, the "one size fits all" imported models of earthquake awareness programs had taken the lives of many people in 2015, as they ran inside looking for safe hideouts under beds or firm tables. The mud walls collapsed and buried many people, as the building infrastructures in Nepal are different from countries where the "one size fits all" earthquake awareness programs are tailor made. Therefore, disaster risk is always a challenge with the lack of cognitive indigenous knowledge, weak institutional and human resource mechanisms for search and rescue, relief, and rehabilitation. The donors blame the recipient state (Pandey, 2009), the state blames the donors, and the communities blame the donors as well as the state's partisan dirty politics and local politics of *bhagbanda* (consensual collusive practices)—an issue that relates closely to the failure to establish strong disaster governance institutions and accountable national and local governments.

The role of effective disaster governance is becoming increasingly important for Nepal, as it is one of the most disaster-prone countries in the world. Anthropogenic political disasters have historically been affecting the day-to-day lives of Nepalese. Regarding natural disasters, every year, many people lose their lives and property. These disasters often badly affect the victims' pathology and psychology for a long time. As Nepal did not have enough capacity, skills, and technologies to fight against the giant earthquake like the disasters in 2015, it appealed to the international community for assistance. There was a downpour of national and international assistance right after the earthquakes. The one and only international airport of Nepal, known as the Tribhuvan International Airport, was in chaos, unable to handle the pressure arising from aid-flights. However, because of the lack of a full-fledged disaster act clearly articulating the "dos and don'ts," there was unwanted chaos in terms of handling search and rescue teams, aid workers, relief materials, distribution systems, and reconstruction processes. At least three issues need to be addressed, if the country wishes to make the reconstruction process successful in a timely manner. First, the overlapping role of sectorial ministries is one of the most important issues that the disaster governance act needs to clarify to avoid institutional disasters. Second, the roles and responsibilities of local NGOs and international NGOs, including other donor organizations, need to be made clear to avoid duplication, redundancy, and other red-tape complications or to avoid institutional and foreign aid disaster (Pandey, 2016). Third, national and local politics that has long been focusing on abstract and normative topics of politics needs to draw its attention seriously back to concrete reconstruction by putting the reconstruction process on the priority list of national and local politics.

NOTES

1. Interview with quake-survived victim in Shankhu on December 5, 2016.
2. Priority number four of the Sendai Framework notes that the recovery, rehabilitation, and reconstruction phase is a critical opportunity to "Build Back Better."
3. Interview December 6, 2016, Earthquake victim representatives from Sindhupalchowk, one of the most affected districts by Nepal earthquakes 2015.
4. Interview December 5, 2016. We interviewed about 15 earthquake victims in Shankhu area on December 5, 2016 to understand the knowledge gap between what the NRA, INGOs, and NGOs were claiming and what people at the ground had received in terms of reliefs and reconstruction.

ACKNOWLEDGMENT

I would like to thank Dr. Dinesh Paudel, assistant professor in the Sustainable Development Department at Appalachian State University, for his review and comments on this manuscript.

REFERENCES

CDRMP. (2011). *Nepal: Comprehensive disaster risk management program.* Retrieved from: http://www.np.undp.org/content/nepal/en/home/operations/projects/environment_and_energy/cdrmp/background.html

Folke, C., Carpenter, S., Elmqvist, T., Gunderson, L., Holling, C.S., & Walker, B. (2002). Resilience and sustainable development: Building adaptive capacity in a world of transformations. *Ambio, 31*(5), 437–440.

Giri, N. (2013). *Implementation of Nepal National Building Code through automated building permit system.* Retrieved from http://www.preventionweb.net/files/35258_35258automationcasestudy.pdf

GoN (Government of Nepal). (2015). *Nepal disaster report 2015*. Kathmandu: MoHA & DPNet-Nepal.

GoN-NPC. (2015). *Nepal earthquake 2015: Post disaster needs assessment*. Kathmandu: National Planning Commission.

Gunderson, L. H., & Holling, C. S. (Eds.). (2002). *Panarchy: Understanding transformations in systems of humans and nature*. Washington, DC: Island Press.

Holling, C. S. (1973). Resilience and stability of ecological systems. *Annual Review of Ecology and Systematics*, *4*, 1–23.

NPC. (2015). *Post disaster needs assessment*. National Planning Commission- Government of Nepal Kathmandu, Nepal.

NRA. (2015). *About national reconstruction authority*. Retrieved from: http://nra.gov.np/pages/details/about

NSDRM. (2008). *National strategy for disaster risk management in Nepal*. Kathmandu: GoN & UNDP. Retrieved from: http://www.nrcs.org/sites/default/files/pro-doc/NSDRM%20Nepal.pdf

Pandey, C. (2016). *Chaotic disaster governance*. Retrieved from: http://reviewnepal.com/articles/chaotic-disaster-governance.html

Pandey, D.R. (2009). Nepal's failed development reflections on the mission and the maladies. Tenth Anniversary ed. Kathmandu, Nepal, Nepal South Asia Centre.

Paudel, K. (2014). Kathmandu valley quake real risk. Retrieved from: http://www.spotlightnepal.com/News/Article/KATHMANDU-VALLEY-QUAKE-Real-Risk

Republica. (2015). Growing complaints forces fake quake victims to return aid money. Retrieved from: http://admin.myrepublica.com/society/story/24189/growing-complaints-forces-fake-quake-victims-to-return-aid-money.html

Republica. 2016. Quake victims to get Rs 300,000 as grant, 500,000 as loan. Available at: http://www.myrepublica.com/news/5304

SWC. (2016). *Update: INGOs brief information*. Kathmandu: social welfare council Nepal. Retrieved from: http://www.swc.org.np/wp-content/uploads/2016/03/List-of-INGOs-2072_073-Falgun-Masant.pdf

Thapaliya, Ram. (2016). An update on reconstruction work in Nepal, at a workshop on '18 months after the Nepal's earthquakes: practical disaster justice in the recovery work' held on 6 December 2016 at Institute of Engineering Nepal. Retrieved from: http://www.sias-southasia.org/past-events/workshop-on-18-months-after-the-nepals-earthquakes-practical-disaster-justice-in-the-recovery-work/

The Guardian-UK. (2015). Nepal government criticized for blocking earthquake aid to remote areas. Retrieved from: https://www.theguardian.com/world/2015/may/02/nepal-government-criticised-blocking-earthquake-aid-remote-areas

The Himalayan Times. (2015). Quake-hit to get first tranche of grant by Sept 18. Retrieved from: https://thehimalayantimes.com/nepal/quake-hit-get-first-tranche-grant-sept-18/

The Himalayan Times. (2016). Distribution of winter cash relief to quake victims on. Retrieved from: http://thehimalayantimes.com/kathmandu/distribution-of-winter-cash-relief-to-quake-victims-on/

The New York Times. (2016). A year after earthquake, Nepal's recovery is just beginning. Retrieved from: http://www.nytimes.com/2016/05/01/world/asia/nepals-earthquake-recovery-remains-in-disarray-a-year-later.html?_r=0

The Rising Nepal. (Undated). Approaching monsoon and relief to quake victims: government must speed up rescue, relief distribution. Retrieved from: http://therisingnepal.org.np/news/3248

The Rising Nepal. 2016. About 416,046 Earthquake-victim households receive first installment of grant. Retrieved from: http://therisingnepal.org.np/news/14840

UN-OCHA. (2015). Nepal earthquake humanitarian response: April to September 2015. Retrieved from: https://www.humanitarianresponse.info/system/files/documents/files/nepal_earthquake_humanitarian_response_report_lr_0.pdf

United Nations. (2015). Sendai framework for disaster risk reduction 2015-2030. New York: United Nations Publication.

Walker, B., C. S. Holling, S. R., Carpenter, & A. Kinzig. (2004) Adaptability and transformability in social-ecological systems. *Ecology and Society*, *9*(2), 5. Retrieved from: https://www.ecologyandsociety.org/vol9/iss2/art5/

Weichselgartner, J., & Kelman, I., 2014. Geographies of resilience: Challenges and opportunities of a descriptive concept. *Progress in Human Geography*, *39*(3), 249–267.

CHAPTER 10

INTERNATIONAL HUMANITARIAN ASSISTANCE AND DISASTER RECOVERY IN ASIA

William L. Waugh, Jr.

ABSTRACT

International humanitarian assistance usually arrives quickly following a catastrophic disaster, although it may be slower to remote locations. The international community has developed guidelines to reduce the social and cultural intrusiveness of the aid, assuring that local priorities are followed and the aid facilitates long-term recovery. However, the aid missions are under pressure to act quickly and withdraw because of the expense of operations, and thus, they are less sensitive to local culture and priorities than they might be. This chapter looks at the political context of international humanitarian assistance, including the Hyogo and Sendai Frameworks and humanitarian standards, and the experience in several catastrophic disaster responses in Asia. Levels of satisfaction with recovery, particularly housing recovery, were related to the affected communities' participation in the decision-making process. Humanitarian aid standards also encourage

Recovering from Catastrophic Disaster in Asia
Community, Environment and Disaster Risk Management, Volume 18, 177–194

ISSN: 2040-7262/doi:10.1108/S2040-726220160000018009

attention to issues of security, displaced populations, equity in the distribution of aid, the safety of women and children, and other disaster impacts.

Keywords: Humanitarian assistance; long-term recovery; Hyogo Framework; United Nations; Nepal earthquake of 2015; Indian Ocean earthquake and tsunami of 2004

INTRODUCTION

International disaster relief is remarkably fast given distances and logistical challenges (Barnett, 2005). The international community generally responds to catastrophic disasters anywhere in the world, particularly when they occur in nations that lack the economic and administrative capacities to respond effectively and recover quickly without aid. The initial obstacle is the willingness of the government in the disaster area to allow foreign governmental and nongovernmental organizations to enter their sovereign territory. There have been recent cases in which governments have declined outside assistance, because they felt that they did not need help or they simply did not want foreign organizations working within their borders. Such was the case in Chile in 2010 when a major earthquake and a resultant tsunami caused serious damage along the coastline. Chile's earthquake-resistant building codes limited the damage to buildings, although serious damage was caused to bridges, roads, and other critical infrastructure. This does not mean that aid is not provided. There also have been cases in which governments have declined international assistance even though help was desperately needed. International involvement can be viewed as intrusive in terms of its impact upon national and local politics, culture, economics, and/or social values.

There is potential for international humanitarian organizations to become involved in civil strife, even if they try to remain neutral, and certainly there is a debate within the international community concerning their proper role in such situations (Stoddard, 2006). International aid workers were attacked by the military in Sri Lanka during that nation's civil war because of concerns that the workers may have been helping rebel forces (Kassenberg, 2017). To be sure, neutrality is not always possible given the nature of some conflicts (Barnett, 2005; Stein, 2005). There are also cases in which countries have simply not been prepared to accept outside aid. Setting the parameters for the delivery of assistance, including setting priorities, can be complicated. In some measures, that was the case in the United States during the Hurricane

Katrina disaster in 2005. It took some time before policies and procedures were in place to handle the international assistance that was offered.

Under the best of circumstances, disaster recovery can be a very long road. The recovery effort in Haiti since the 2010 earthquake, for example, has been frustrating for residents and for foreign development and humanitarian organizations alike. Political turmoil has made it difficult to design and implement policies, and the lack of administrative capacity in the country has confounded efforts to manage programs. After 2 years and US$4.1 billion pledged, Nepal's recovery from the April 2015 earthquake has been very slow. The quake killed 9,000 and destroyed over 824,000 homes (Adhikari, 2017). Mountain villages were particularly hard hit. The government is now trying to shut down camps housing the homeless, many of whom came from mountain villages and are afraid to return for fear of another earthquake. Politics and a lack of administrative capacity have held up the transfer of money to survivors to rebuild their homes.

International humanitarian assistance has changed over the past several decades. The International Decade for Natural Hazard Reduction in the 1990s encouraged states to adopt policies and programs to reduce risks and better prepare for natural disasters. Through the International Strategy for Hazard Reduction and a series of international agreements, risk reduction has become a major international effort and that effort has influenced the delivery of humanitarian assistance by the United Nations (UN) agencies and most governmental and nongovernmental aid programs. The World Food Program, for example, provides a limited menu that can be tailored to cultural food preferences. The objective of this analysis is to examine the international context of humanitarian assistance and the relationship between international humanitarian assistance and recovery from catastrophic disasters in Asia. The methodology for gauging need following catastrophic disasters (see Stoddard & Hoffman, 2003) has become more sophisticated, and strong efforts are being made to forecast need well before a major disaster, including estimating thresholds of need based upon local diet, mortality, and other factors and the requirements to address dietary shortfalls.

International humanitarian organizations and their major donors also bring their own values to aid missions. Social and political equality, women's rights, anti-corruption, and other values can drive international aid and can become conditions for receiving aid. Donors often contribute money and other resources, because they have particular interests, such as schools, children, agriculture, or business recovery, or certain values that they wish to promote. CARE USA, for example, supports women and girls, including the provision of separate toilets to reduce the potential for violence and local capacity building (Kassenberg, 2017). Although there are ethical constraints on religious organizations using aid missions to proselytize, there may be fewer constraints on encouraging other social and political values. Certainly some religious organizations do send aid to support affiliated churches and missions. Within the UN and humanitarian networks, there are guidelines and principles to ensure noninterference with host governments. InterAction, a coordinating body for international relief organizations, has guidelines for its members that begin with the following statement: "A member's programs shall respect the dignity, values, history, religion and culture of the people served" (see InterAction.org, PVO Standard 7.1.5). Member policies and practice should be nondiscriminatory, based upon need, and "never be preconditioned upon faith ... or tied to the embrace or acceptance of a particular political or religious creed."

What we think of as the international humanitarian assistance community includes international organizations like the UN agencies that deal with refugees and aid, economic alliances, national foreign aid agencies, international development organizations, secular and religious organizations, professional associations, ad hoc aid organizations created for specific disasters, private sector organizations, and volunteers—to mention but a few possibilities. Aid might also be provided by foreign military personnel, particularly when the security of affected populations and/or relief personnel is a major concern (O'Callaghan & Pantuliano, 2007), and humanitarian aid workers are being trained to recognize threats to themselves, as well as to the populations affected by disaster (Kassenberg, 2017). During the Haitian earthquake disaster in 2010, military units were paired with relief organizations to transport and distribute food and medicine to ensure the security of personnel and material. Development and charitable agencies working in Haiti became humanitarian assistance organizations by virtue of their involvement in the disaster response and recovery and their connections to Haitian institutions. In other words, the lines between humanitarian assistance and development have become somewhat blurred.

The language and practice of international humanitarian assistance and domestic disaster management are also sounding and looking more and more alike. In both cases, the organizations are becoming more and more professional, and the training and educational programs are overlapping more and more. Today, humanitarian assistance and emergency management require experience and specialized training. Management skills, interpersonal skills, cultural sensitivity, language skills, and some technological skills are increasingly necessary. Well-meaning, but untrained, relief workers are becoming rare. Cultural competency can be critical in working with survivors and local officials (Southall, DeYoung, & Harris, 2017) and cultural interoperability can be critical working with other organizations, including other humanitarian relief organizations (Waugh, 2002, 2007). Sensitivity to the ethical dimensions of their work is also necessary and today that means recognition that humanitarian assistance should fit local response and recovery priorities, not conflict with local custom and culture, and it should contribute to the recovery process. The literature increasingly refers to sustainable assistance or helping communities rebuild, reconstruct, and redevelop in a manner that is sustainable (Waugh & Liu, 2014).

THE INTERNATIONAL CONTEXT

The Hyogo Framework was developed in January 2005 in Kobe, Hyogo, Japan—with 168 governments adopting a 10-year plan to make the world safer from natural hazards. The global strategy is to engage a wide range of actors in a coordinated effort to reduce the risks of disasters and to build a "culture of prevention" in society as part of sustainable development. Out of that Framework a network of national platforms was developed (see Matsuoka & Shaw, 2014).

In essence, the Hyogo Framework was initiated to ensure that disaster risk reduction is a national and local priority. The goal is to cultivate a strong foundation for the implementation programs to

- Identify, assess, and monitor disaster risks and enhance early warning
- Use knowledge, innovation, and education to build a culture of safety and resilience at all levels
- Reduce the underlying risk factors
- Strengthen disaster preparedness for effective response at all levels
- Community participation
- State investment
- Integration into development
- Multihazard approach
- Capacity development
- Decentralized responsibility
- Gender as a core factor
- Public–private partnerships

National platforms are necessary to tailor the programs to national circumstances and needs with appropriate financial, administrative, and regulatory resources being applied, for example, budgets or staff time, goals, and impacts. Reducing vulnerability is the overall goal. Measures of achievement include reducing losses, integrating goals into development programs, integration into emergency plans, greater disaster preparedness, risk reduction, and development of a "culture of safety."

The Sendai Framework for Disaster Risk Reduction 2015–2030 improved on the Hyogo Framework by providing targets on the basis of the standards (above) and an instrument for measuring progress (UNISDR, 2015). Efforts are underway to develop technological and scientific tools with which to measure progress toward those goals.

HUMANITARIAN ASSISTANCE

The strengthening of national and international support for hazard reduction has affected the conduct of humanitarian operations. That can be seen in the emphases on local capacity-building, community participation, gender issues, and linkages between hazard mitigation and development. Local priorities are paramount. This does not always mean that local priorities are followed because of the pressure to provide shelter, food, and other necessities quickly. It does mean that the explicit goals include attention to local priorities. Conflict between the goals of the humanitarian assistance organizations and their clients in disaster areas can be fundamental, but there are still guidelines that support the primacy of local priorities.

The Cluster System was developed to organize international relief efforts by sector with, for example, the World Food Program being the lead for both food aid and logistics. However, the initial assessment of the approach following the Pakistan earthquakes in 2005 concluded that it improved the process for coordinating the efforts of multiple sectors and identifying and addressing gaps in programming. The identification of lead agencies also clarified roles and leadership. Some relationships with NGOs were improved. Goal setting and efficiency were improved, but relationships with host nations and global performance were mixed. Clusters tended to remain inclusive with minimal interaction with local government officials and organizations (Stoddard et al., 2007, pp. 1–2). Indeed, meetings might be conducted in a language unfamiliar to locals, and it was uncertain that local needs were being addressed.

For example, the Humanitarian Response Index 2010 outlined guidelines for international humanitarian assistance, including whether responses are (1) based on the needs of affected populations, rather than other outside interests; (2) strengthen local capacity and long-term recovery; (3) policies and programs support humanitarian organizations; (4) follow and support international humanitarian law and the protection of affected civilian populations; and (5) contribute to learning and accountability in humanitarian action. These are referred to as the five pillars. The focus is attention to local needs.

The scores are based on the following:

Pillar 1—Responding to needs (30% of index weight)
Pillar 2—Prevention, risk reduction, and recovery (20% of index weight)
Pillar 3—Working with humanitarian partners (20% of index weight)
Pillar 4—Protection and international law (15% of index weight)
Pillar 5—Learning and accountability (15% of index weight)

The five main findings in the 2010 assessment were as follows:

1. Humanitarian assistance was increasingly politicized and "millions of people are not getting the aid they need."
2. Gaps in the protection of civilians and the lack of safe access for humanitarian organizations means that civilian populations are at risk.
3. Donors lack commitment to disaster and conflict prevention, including commitment to addressing vulnerabilities to climate change.
4. The humanitarian system has only slowly reformed and greater attention needs to be paid to improvements to increase efficiency and effectiveness.
5. Improvements need to be made in transparency and accountability to affected populations. The overall ranking shows that there is still great room for improvement in the way all donor governments apply Good Humanitarian Donorship (GHD) principles in their policies and practices. Overall, donors did "reasonably well" in Pillar 1, "less well" in Pillar 2, mixed in Pillar 3, "reasonably consistent" in Pillar 4, and very mixed in Pillar 5. The assessment went on to recommend that donors address the five pillars better. The Humanitarian Response Index is linked to the GHD Initiative (see Saban, 2016, for summaries of major humanitarian codes of ethics and practice).

LINKING INTERNATIONAL DISASTER RELIEF AND LOCAL RECOVERY

The direct impact of international relief on recovery is less clear than it might be. It must also be said that there are no agreed upon methodologies or indicators to assess humanitarian assistance impacts themselves (Moslehi, Ardalan, Tirone, Waugh, & Akbarisari, 2016; Moslehi, Ardalan, Waugh,

Tirone, & Akbarisari, 2016). Generally, when external aid arrives, recovery has already begun. Indeed, disaster responses have begun before trained emergency responders arrive. Family, friends, and neighbors often have begun search and rescue operations and provided some levels of emergency medical care for those injured. Community-level responses and perhaps some regional and national responses have also begun. The "first responders" are generally neighbors, friends, and family, rather than professional responders. Humanitarian relief organizations may well be gone before recovery is reasonably complete.

For international humanitarian organizations, relief and reconstruction processes are driven by pressure to spend donors' money and achieve quick results. There is a tendency to overlook local culture, social networks, local capacities, and local needs, despite the principles that generally guide relief operations. Because of the pressure to respond quickly, humanitarian organizations tend to deviate from their own plans and from the imperative to work closely with affected communities. There are a lot of reasons for the deviation. Money is one reason. Funding is limited and work has to be carried out quickly and, often, by contractors who work by the hour, day, week, or month. The logistics involved in delivering humanitarian aid is expensive and can be prohibitive. Building materials may be scarce and contractors have to use what they can acquire locally. Many international and domestic disaster relief organizations ask for contributions of money instead of food, clothing, and other materials, because money is easily transferred and can be reprogrammed as needs are identified or change. When reconstruction and recovery are complete, their time runs out or their money runs out, and they pack up and leave (Barenstein & Leeman, 2013). Time and resources are finite and deployments of personnel are generally limited. In essence, relief organizations cannot stay engaged for extended periods. They have to respond to new disasters.

Community participation is time consuming and there may or may not be a cohesive or even identifiable "community" in place. If local governance structures are still working and local leaders can identify recovery priorities, the recovery process is much more likely to be successful (Alesch, Arendt, & Holly, 2009). In the rebuilding of Aceh after the 2004 earthquake and tsunami, local satisfaction with the recovery process was related to the reconstruction approach of the humanitarian organization, the participation of local leaders in the management of the reconstruction processes, and the accountability of the leader to the community (Mantel, 2013). In Gujarat, India, after the 2001 earthquake, satisfaction levels were highest with the owner-driven approach supported by government and the work was faster and more cost effective. When it was an agency-driven approach with limited owner participation, satisfaction levels were lower. Local NGOs tended to abandon development principles, because they were under pressure to finish quickly (Barenstein, 2013a; Barenstein & Pittet, 2013). Gender roles were also affected in Gujarat by changes in housing technologies (Barenstein, 2013b). Ultimately, new homes may not fit local living conditions and preferences. In Tamil Nadu, contractors cut down trees as they were building homes and, in doing so, removed what locals considered critical living space under the trees. Community space was lost (Naimi-Gasser, 2013).

Linking redevelopment to sustainability is difficult within the disaster recovery timeframe and within the constraints of local capacities (Barenstein & Leeman, 2013). Poor governance and corruption can reduce the impact of disaster aid and divert resources from the long-term recovery effort (Barone & Mocetti, 2014). Choices are made early on that preclude transition to new patterns of economic activity and social interaction. The fact of the matter is that some communities recover and some do not.

Aldrich (2012) argues that the key variable is the presence of social capital. Social capital serves as informal insurance and mutual assistance and, through collective action, makes it possible to do things that individuals and small families cannot. Social capital, including community networks, also encourages disaster survivors to stay in or return to their communities (p. 2). There is greater tendency to leave and not return if survivors are less connected to their communities. As noted earlier, "community" is a relative term and just because someone lives within a village or town or city does not mean that they are connected with neighbors.

As stated by Aldrich, "...community recovery is the process of repopulation by survivors – who may have fled or been evacuated – and new residents along with the gradual resumption of normal daily routines for those occupants" (p. 5). He looks for population change, receipt of aid packages, rebuilding (construction), and occupation of temporary housing as indicators of recovery. If the population remains the same or nearly so, outside aid is no longer needed, reconstruction is progressing, and fewer people are in temporary housing, recovery is occurring. In the United States, attention focused on the numbers of evacuees who did not return to New Orleans and other affected areas, the in-migration of people involved in the reconstruction of the city and surrounding area (such as Hispanics), and the investment in reconstruction.

Quality of governance and leadership are often the keys in disaster recovery (Aldrich, 2013; Alesch et al., 2009). The poor responses to the Hurricane Katrina disaster in the United States (Waugh, 2006) and the Kobe earthquake in Japan are cases in point. Nonetheless, some communities recover quickly because of the quality of their leadership or even despite their leaders. Aid money can also have perverse effected, such as elevated dowries in India and increased divorce in New York, because survivors could afford to pay alimony and lawyers. Housing issues can also arise, because poor residents often do not have legal ownership of the land on which their homes and businesses sit. Following the earthquake and tsunami in 2004, poor residents tended to be pushed away from the beaches in Thailand, because the hotels and restaurants wished greater access for patrons. That practice has been common in developed and underdeveloped nations. The poor have few housing options. The poor are also more likely to lose their livelihoods. Following the Hurricane Katrina disaster, some businesses began providing temporary housing for employees to ensure that they would be available once the businesses were back in operation. Racism and nativism can also affect the allocation of disaster aid from governments, nongovernmental organizations, and private firms.

The factors that become obstacles to recovery include population density. Public housing, slums, and other high-density, low-income areas become problematic as housing demand is high, but there are few incentives for developers and development agencies to build such housing. Social capital translates into better health, stronger community governance, civic engagement, trust in government, economic growth, and resilience. Marginalized groups may not benefit and they may become targets of violence (Aldrich, 2012).

Aldrich's recovery variables were (Aldrich, 2014, p. 15):

- Quality of governance
- External aid
- Amount of damage
- Population density
- Demographics/socioeconomic conditions
- Social capital

According to Aldrich, "high levels of social capital – more than such commonly referenced factors as socioeconomic conditions, population density, amount of damage or aid – serve as the core engine of recovery" (Aldrich, 2012, p. 15) His case studies were the Tokyo earthquake of 1923, the Kobe earthquake of 1995, the Indian Ocean tsunami of 2004, and Hurricane Katrina in 2005. His conclusions support the idea that social capital is an important resource for recovery. Social networks bond communities together. Social networks can bridge communities within a nation and across borders.

And they can link social networks together. The bonds provide buy-in to facilitate collective action and mutual assistance. They can also create distrust of outsiders and an unwillingness to work with those outside of the group and, hence, the opposition to mobile homes in New Orleans and violence against ethnic Koreans during the 1923 Tokyo earthquake response and recovery. Strong social networks can dominate.

International humanitarian assistance was very limited in 1923. A tsunami of aid followed the Great Tsunami of 2004. Over 230,000 people were killed in Indonesia, India, Sri Lanka, and Thailand. One was killed in Africa. In Tamil Nadu, India, the greatest toll was among women and children because many of the men were fishermen and were not affected by the waves. Warnings may have reached some communities. Parish councils, urr panchayats (fishermen's associations), and hamlet councils made decisions on the allocation of assistance. There were penalties for not turning in aid received from outside, and the subsequent allocations were on the basis of preferences that generally excluded lower caste people, widows, single women, and other marginal peoples. Aldrich relates the story of how one council collected donated fishing boats and sold them with the money being distributed to members of the community, rather than letting poor fisherman have them as the donors intended. Similarly, other donations were turned in and given to others. The point is that discrimination was common. Those with connections in the central government, nongovernmental organizations, and/or outside of India fared better. International assistance became the "second tsunami." Relief organizations also preferred to build single-family homes despite the fact that the home they were expected to replace were for extended families. The targeting of humanitarian assistance was poor, and some of the worst damaged communities received little or no aid. Local culture and social relations suffered.

Aldrich's findings, as well as the findings of other studies of social networks, have raised questions about how to maintain social networks during evacuation and temporary sheltering. More fundamentally, can officials encourage the development and strengthening of social networks to make communities more resilient? Social capital is the secret to community resilience, according to Aldrich (2017). The relationships and support act to give residents a sense of efficacy and provide the foundation for collective action. In the Tamil Nadu case, poor planning resulted in some survivors remaining in displacement camps even when they could have returned home. More overall planning could have alleviated that problem. The absence of baseline data on communities made it difficult to assess the needs and target.

To improve coherence, effectiveness, and accountability, the GHD Initiative was developed in 2003. The initiative "called for humanitarian assistance to be provided in such a way that it could support recovery and long-term development, help to maintain and restore sustainable livelihoods and support the transition from humanitarian relief to recovery and development." The Paris Declaration in 2005 identified five principles of humanitarian aid effectiveness, and the Accra Agenda for Action called for a greater multidimensional approach to speed recovery and local capacity-building to "build back better."

Sustainable recovery is the goal. "Building back better," however, raises a number of issues relative to the determination of what is and who chooses what is "better." In an environment in which quick action is necessary to shelter, feed, and provide medical care for those affected by disaster, sustainable recovery generally is less of an imperative than life-saving activities. Mass feeding, mass sheltering, and other life-saving activities may be all that can be expected with limited funds and time pressure. There are also ethical issues related to making long-term choices for communities when they should be making their own recovery decisions. The underlying propositions for the "building back better" approach were outlined by William J. Clinton, Special Envoy for the UN Secretary General, on the 2004 Aceh Earthquake and Tsunami relief effort. The propositions were (Clinton, 2006)

1. Governments, donors, and aid agencies must recognize that families and communities drive their own recovery.
2. Recovery must promote fairness and equity.
3. Governments must enhance preparedness for future disasters. Proposition 4: Local governments must be empowered to manage recovery efforts, and donors must devote greater resources to strengthening government recovery institutions, especially at the local level.
4. Good recovery planning and effective coordination depend on good information.
5. The UN, World Bank, and other multilateral agencies must clarify their roles and relationships, especially in addressing the early stage of a recovery process. (Proposition 7: The expanding role of NGOs and the Red Cross/Red Crescent Movement carries greater responsibilities for quality in recovery efforts.)
6. From the start of recovery operations, governments and aid agencies must create the conditions for entrepreneurs to flourish.
7. Beneficiaries deserve the kind of agency partnerships that move beyond rivalry and unhealthy competition.
8. Good recovery must leave communities safer by reducing risks and building resilience.

A 9.1 earthquake struck 240 kilometers from the coast of Aceh Province in Indonesia on December 26, 2004, causing tsunami waves that devastated communities and destroyed facilities in Aceh and other coastal areas of Indonesia, Thailand, Sri Lanka, and elsewhere around the Indian Ocean. Approximately 230,000 people were killed in 14 nations and millions were left homeless. It was one of the worst natural disasters in recorded history. The Indonesian government operation was managed by the National Coordinating Board for Disaster Management within the Office of the Vice-President. The UN was asked to help with the management of international humanitarian assistance and a joint management center was created. The Indonesian military was engaged in search and rescue and burying the dead. Sixteen nations sent approximately 4,500 personnel to assist. By June 2005, six months after the tsunami, US$4.9 billion in rehabilitation and reconstruction projects were underway, according to the World Bank. An additional US$3.1 billion had been pledged (Fan, 2013).

The challenge was to "build back better" and assure a sustainable recovery. Cash transfers were made to some 18,000 program participants. A program to resettle survivors on titled land with joint ownership between husbands and wives with inheritance rights for women and children was initiated in state and sharia courts. Local communities were empowered to control their own priorities (Fan, 2013).

Peace between the national government and GAM (Free Aceh Movement) and between the central government and local authorities were encouraged. Transparency, accountability, and effective service delivery were priorities. Other priorities included encouraging community participation and women's empowerment. The incentives for buy-in were "additional" resources. Aside from rebuilding housing, humanitarian organizations viewed the issue more in terms of empowering local communities, rather than changing the infrastructure (Fan, 2013).

When Cyclone Nargis made landfall in Myanmar in May 2008, it caused catastrophic destruction. Nargis ranked the second deadliest tropical storm in history. The death toll was estimated at 140,000 by the Myanmar government with approximately 20,000 injured. Fishing and farming communities were destroyed. Infrastructure and approximately 700,000 homes were seriously damaged, leaving 800,000 homeless. Within the Ayeyarwady Delta and Yangon region were the hardest hit areas of the country with some 2.8 million inhabitants, and over 75% of the hospitals and clinics, roads and bridges, and power lines were destroyed (Fan, 2013).

Poor relations between the Myanmar military government and the international community made the government reluctant to ask for outside help. Nonetheless, international aid arrived quickly. Access to the disaster area was limited. The United Nations Security Council considered ignoring the Myanmar government's limit on access and assuming responsibility to act in order to save lives, but that course was vetoed by China and the United Kingdom—and the UN Emergency Relief Coordinator, Sir John Holmes. Myanmar relented and formally asked for international humanitarian. An ESEAN Emergency Rapid Assessment Team was accepted and medical teams were deployed (Fan, 2013).

The Nargis response led to the development of networks extending across national boundaries, rural–urban divides, linkages between rural and urban networks, After a year, only US$100 million was raised, less than 20% of the funding sought, because of the distrust of the Myanmar government. Sheltering the displaced was not a priority and, 3 years after Nargis, thousands were still living in shelters that offered little protection from tropical storms. Some were still living in temporary shelters. Donors were reluctant to commit to reconstruction because of the military government (Fan, 2013).

Building back better was an important consideration in both the Aceh and Myanmar cases (as well as the Haiti case in Fan's analysis). However, "build back better" meant different things with each case (Fan, 2013). "Build back better" is a common goal in national risk-reduction programs, although the words may be slightly different. The term may be too vague to guide policy and program. Might the concept be made more concrete so that it can guide action and mobilize support (Fan, 2013)?

CONCLUSIONS

Humanitarian assistance missions in Asia have been very much like those in other parts of the world. However, population density may be greater in Asia, and thus, might increase the number of vulnerable populations and the likelihood of epidemics and even pandemics. Proximity to the Ring of Fire increases risk of earthquakes, tsunamis, and volcanic eruptions. Social networks and social capital might be strong in some nations, as Aldrich found, but that is not a guarantee of community resilience.

Over time, humanitarian aid has become less intrusive in terms of conflicts with local values, customs, and priorities, but there are still strong pressures to get in and get out as quickly as possible and community participation and leadership are time consuming. To the extent that humanitarian operations use third parties, that is, contractors, the goal of local control gets more difficult to achieve. Contractors are not responsible to the local community, they are responsible to the humanitarian organization. Fortunately, some of the cases cited are dated. Humanitarian assistance has changed considerably since the 2001 Gujarat earthquake and even since the 2004 Indian Ocean earthquake and tsunami. Lessons were learned, albeit sometimes painfully.

The professionalization of international humanitarian organizations has meant that such organizations are duty-bound to recruit and train personnel for technical and social skills. There are basic skills related directly to managing field operations (see Cahill, 2003, 2007), but training and education have broadened the focus of international operations. There is greater attention to social–cultural skills, including sensitivity to cultural differences. There is greater attention paid to issues that often accompany international and domestic disaster operations, including human trafficking, violence against women and children, corruption, and inequities in the distribution of aid. Capacity issues are also common to both. Increasingly, governments and organizations of all sorts are being encouraged to engage in pre-disaster recovery planning (Waugh, 2017). The value of identifying housing needs and the aforementioned dietary needs is an important consideration in terms of planning for new housing, food programs, and even deciding how well the local economy and labor force can be relied upon in the recovery effort. Such pre-disaster planning can also provide baseline community data for damage assessment to assess local needs. Cash-based aid offers important economic stimuli, as well as opportunity to engage survivors in operations that can make them feel more efficacious. It is a way to turn victims into survivors and help them drive their own development. Pre-disaster planning may also provide the groundwork for an examination of the trade-offs involved in "build back better" efforts. Critical needs that have to be addressed as quickly as possible can be identified and time can be taken to plan for "better" choices after that. That might also help international relief organizations, and host governments target aid on the greatest needs and design a more equitable distribution of aid. The tough political questions can be recognized even if they are not addressed, and the international community can determine whether and/or how they wish to work with central government and community officials, as well as local nongovernmental organizations.

To the extent that the Hyogo and Sendai Frameworks are encouraging improvement in national risk reduction, as well as disaster response and recovery, and the development of standards promises to increase adherence to humanitarian values, governments will be more capable of managing the disasters that do occur. Catastrophic disasters, however, will continue to overwhelm many nations, and humanitarian relief will be necessary in Asia and elsewhere. However, the infrequency and scale of such events complicates planning and operations. Increasing community resilience will help, particularly in locations prone to catastrophic disasters. Governments also need to prepare for outside assistance in terms of determining what kinds of assistance they might need and how it should be delivered. Developing working relationships with international humanitarian aid organizations is a good idea, and the best way is to invite them into planning processes and disaster exercises. Just as domestic disaster management is facilitated by joint planning and exercises, international disaster management, including long-term recovery, can also be facilitated.

REFERENCES

Adhikari, N. (2017). Nepal's earthquake disaster: Two years and $4.1 billion later. Aljazeera, April 2017. Retrieved from: http://www.aljazeera.com/indepth/opinion/2017/04/nepal-earthquake-disaster-years-41bn-170412110550808.html

Aldrich, D. J. (2012). *Building resilience: Social capital in post-disaster recovery.* Chicago, IL: University of Chicago Press.

Aldrich, D. J. (2017). The importance of social capital in building community resilience. In W. Yan and W. Galloway (Eds.), *Rethinking resilience, adaptation and transformation in a time of change*. Springer International Publishing.

Alesch, D., Arendt, L., and Holly, J. (2009). *Managing for long-term community recovery in the aftermath of disaster*. Fairfax, VA: Public Entity Risk Institute.

Barenstein, J. D., & Leeman (Eds.). (2013). *Post-disaster reconstruction and change: Communities' perspectives*. Boca Raton, FL: CRC Press.

Barenstein, J. D. (2013a). Communities' Perspectives on Housing Reconstruction in Gujarat Following the Earthquake of 2001. In J. D. Barenstein & Leeman (Eds.), *Post-disaster reconstruction and change: Communities' perspectives* (pp. 71–100). Boca Raton, FL: CRC Press.

Barenstein, J. D. (2013b). A social and environmental assessment of pre- and post-tsunami housing and building practices in Tamil Nadu. In J. D. Barenstein & Leeman (Eds.), *Post-disaster reconstruction and change: Communities' perspectives* (pp. 119–136). Boca Raton, FL: CRC Press.

Barnett, M. (2005). Humanitarianism transformed. *Perspectives on Politics*, *3*(4) December, 723–740.

Barone, G., & Sauro, M. (2014). Natural disasters, growth and institutions: A tale of two earthquakes. *Journal of Urban Economics*, *84*, 52–66.

Cahill, K. M. (Ed.). (2003). *Basics of international humanitarian missions.* New York: Fordham University Press.

Cahill, K. M. (2007). *The pulse of humanitarian assistance.* New York: Fordham University Press.

Christopher, M., & Tatham, P. (Eds.). (2011). *Humanitarian logistics: Meeting the challenge of preparing for and responding to disasters*. UK: Kogan Page Ltd./The Chartered Institute of Logistics and Transport.

Clinton, W. J. (2006). *Key propositions for building back better: A report by the UN Secretary-General's Special Envoy for Tsunami Recovery*. William J. Clinton, United Nations, December.

Darcy, J., & Hoffman, C.-A. (2003). *According to need? Needs assessment and decision-making in the humanitarian sector*. Overseas Development Institute, HPG Report 15, September.

Fan, L. (2013). *Disaster as opportunity? Building back better in Aceh, Myanmar, and Haiti*, HPG Working Paper, November 2013.

Glaser, M. P. (2005). *Humanitarian engagement with non-state armed actors: The parameters of negotiated access*. Overseas Development Institute, HPG Report 51, June.

Hannigan, J. (2012). *Disasters without borders: The international politics of natural disasters*. Cambridge, UK: Polity Press.

InterAction. (2015). *PVO Standard 7.1.5*. Retrieved from: www.interaction.org, accessed 15 November.

Kassenberg, L. (2017). The roles of NGOs in disaster management and response. Webinar: The roles of NGOs in disaster management and response. William Averette Anderson Fund, May 9.

Mantel, C. (2013). Ownership, control, and accountability in post-tsunami housing reconstruction processes in Aceh, Indonesia. In J. D. Barenstein & Leeman (Eds.), *Post-disaster reconstruction and change: Communities' perspectives* (pp. 55–70). Boca Raton, FL: CRC Press.

Matsuoka, Y. & Shaw, R. (2014). *Hyogo framework for action and urban disaster resilience*. UK: Emerald Group Publishing Ltd.

Moslehi, S., Ardalan, A., Waugh, W. L., Tirone, D. C., & Akbarisari, A. (2016). Characteristics of an effective international humanitarian assistance: A systematic review. *PLOS: Currents – Disasters*, February 25.

Moslihi, S., Ardalan, A., Tirone, D. C., Waugh, W. L., Jr., & Sari, A. A. (2016). Procedural effectiveness of international humanitarian assistance. *Global Journal of Health Science*, *8*(11).

Naimi-Gasser, J. (2013). The remembered trees: Contractor-driven reconstruction and its consequences on communities' well-being in coastal Tamil Nadu. In J. D. Barenstein & Leeman (Eds.), *Post-disaster reconstruction and change: Communities' perspectives* (pp. 137–155). Boca Raton, FL: CRC Press.

O'Callaghan, S. & Pantuliano, S. (2007). *Protective action: Incorporating civilian protection into humanitarian response.* Overseas Development Institute, Humanitarian Policy Group, Report 26, December.

Saban, L. I. (2016). *International disaster management ethics*. Albany, NY: SUNY Press.

Smillie, I., & Minear, L. (2004). *The charity of nations: Humanitarian action in a calculating world.* Kumarian Press.

Southall, H. G., DeYoung, S., & Harris, C. A. (2017). Lack of cultural competency in international aid responses: The Ebola Outbreak in Liberia, *Frontiers in Public Health*, *5*, January 2017. Retrieved from: www.frontiersin.org, January 31, 2017.

Stein, J. (2005). Humanitarianism as political fusion. *Perspectives on Politics*, *3*(4), 741–744.

Stoddard, A. (2006). Providing aid in insecure environments: Trends in policy and operations. *Humanitarian Policy Group Briefing Paper*, *24*, September.

Stoddard, A., Harmer, A., Haver, K., Salomons, D., & Wheeler, V. (2007). *Cluster approach evaluation: Final report*. Overseas Development Institute, Humanitarian Policy Group. November.

United Nations Office for Disaster Risk Reduction. (2015). *Sendai framework for disaster risk reduction 2015–2030*. Geneva, Switzerland: UNISDR.

Waugh, W. L., Jr. (2002). Organizational culture, communication, and decision-making: Making multi-organizational, inter-sector and intergovernmental operations work. In *2002 National Conference on Catastrophic Care for the Nation, National Disaster Medical System*. Atlanta, GA, April 13–17.

Waugh, W. L., Jr. (2007). Hurricane Katrina and cultural interoperability. In *American Society for Public Administration National Conference*. Washington, DC, March 23–27.

Waugh, W. L., Jr., Liu, C. Y. (2014). Disaster, the whole community, and development as capacity-building. In Naim K. (Ed.), *Disasters and development: Conceptual issues and theoretical perspectives*. Springer, pp. 167–180.

Waugh, W. L., Jr. (2017). The value of pre-disaster recovery planning for post-disaster recovery. In *78th National Conference of the American Society for Public Administration*. Atlanta, Georgia, March 17–22.

Welton-Mitchell, C., Korbel, J., James, L., Awale, R. (2016). Nepal 2015 earthquake: A rapid assessment of cultural, psychological and social factors with implications for recovery and disaster preparedness. In *International Journal of Mass Emergencies and Disasters*, *34*(3), 399–418.

INDEX